Review of the Lake Ontario–St. Lawrence River Studies

Committee to Review the Lake Ontario-St. Lawrence River Studies

Water Science and Technology Board

Division on Earth and Life Studies

NATIONAL RESEARCH COUNCIL
OF THE NATIONAL ACADEMIES

in collaboration with

The Royal Society of Canada

THE NATIONAL ACADEMIES PRESS
Washington, D.C.
www.nap.edu

THE NATIONAL ACADEMIES PRESS 500 Fifth Street, N.W. Washington, DC 20001

NOTICE: The project that is the subject of this report was approved by the Governing Board of the National Research Council, whose members are drawn from the councils of the National Academy of Sciences, the National Academy of Engineering, and the Institute of Medicine. The members of the committee responsible for the report were chosen for their special competences and with regard for appropriate balance.

Support for this project was provided by the Department of State under award number SAQMPD05C1133. Any opinions, findings, conclusions, or recommendations expressed in this publication are those of the author(s) and do not necessarily reflect the views of the organizations or agencies that provided support for the project.

International Standard Book Number 0-309-10068-2 (Book)

Additional copies of this report are available from the National Academies Press, 500 Fifth Street, N.W., Lockbox 285, Washington, DC 20055; (800) 624-6242 or (202) 334-3313 (in the Washington metropolitan area); Internet, http://www.nap.edu.

Printed in the United States of America.

THE NATIONAL ACADEMIES

Advisers to the Nation on Science, Engineering, and Medicine

The **National Academy of Sciences** is a private, nonprofit, self-perpetuating society of distinguished scholars engaged in scientific and engineering research, dedicated to the furtherance of science and technology and to their use for the general welfare. Upon the authority of the charter granted to it by the Congress in 1863, the Academy has a mandate that requires it to advise the federal government on scientific and technical matters. Dr. Ralph J. Cicerone is president of the National Academy of Sciences.

The **National Academy of Engineering** was established in 1964, under the charter of the National Academy of Sciences, as a parallel organization of outstanding engineers. It is autonomous in its administration and in the selection of its members, sharing with the National Academy of Sciences the responsibility for advising the federal government. The National Academy of Engineering also sponsors engineering programs aimed at meeting national needs, encourages education and research, and recognizes the superior achievement of engineers. Dr. Wm. A. Wulf is president of the National Academy of Engineering.

The **Institute of Medicine** was established in 1970 by the National Academy of Sciences to secure the services of eminent members of appropriate professions in the examination of policy matters pertaining to the health of the public. The Institute acts under the responsibility given to the National Academy of Sciences by its congressional charter to be an adviser to the federal government and, upon its own initiative, to identify issues of medical care, research, and education. Dr. Harvey V. Fineberg is president of the Institute of Medicine.

The **National Research Council** was organized by the National Academy of Sciences in 1916 to associate the broad community of science and technology with the Academy's purposes of furthering knowledge and advising the federal government. Functioning in accordance with general policies determined by the Academy, the Council has become the principal operating agency of both the National Academy of Sciences and the National Academy of Engineering in providing services to the government, the public, and the scientific and engineering communities. The Council is administered jointly by both Academies and the Institute of Medicine. Dr. Ralph, J. Cicerone and Dr. Wm. A. Wulf are chair and vice-chair, respectively, of the National Research Council.

www.national-academies.org

The **Royal Society of Canada**, *The Canadian Academy of the Sciences and Humanities*, is the senior national body of distinguished Canadian scientists and scholars. Its primary objective is to promote learning and research in the arts and sciences. The Society consists of men and women from across Canada who are selected by their peers for outstanding contributions to the natural and social sciences and in the humanities. Dr. Gilles Paquet is president of the Royal Society of Canada.

COMMITTEE TO REVIEW THE LAKE ONTARIO-ST. LAWRENCE RIVER STUDIES*

JAMES L. WESCOAT, *Chair*, University of Illinois, Urbana-Champaign
PATRICIA CHOW-FRASER, McMaster University, Hamilton, Ontario, Canada
DESMOND N.D. HARTFORD, British Columbia Hydropower Authority, Burnaby, Canada
JANET R. KEOUGH, U.S. Environmental Protection Agency, Duluth, Minnesota
LYNN ALISON MAGUIRE, Duke University, Durham, North Carolina
JOSEPH C. MAKAREWICZ, State University of New York at Brockport, New York
DAENE C. MCKINNEY, University of Texas, Austin
DAVID H. MOREAU, University of North Carolina, Chapel Hill
SLOBODAN P. SIMONOVIC, University of Western Ontario, London, Canada
P. ANDY ZIELINSKI, Ontario Power Generation, Toronto, Canada

National Research Council Staff

LAUREN E. ALEXANDER, Study Director
ELLEN A. DE GUZMAN, Research Associate

*The activities of this committee were overseen and supported by the U.S. National Research Council's Water Science and Technology Board (WSTB) in collaboration with the Royal Society of Canada (RSC). The WSTB members are listed in Appendix D. Dr. Jeremy N. McNeil, chair of the RSC committee on expert panels, served as liaison from the Royal Society. Biographical information on committee members is contained in Appendix E.

Preface

This report reviews selected works of the Lake Ontario-St. Lawrence River Study for the International Joint Commission (IJC). The IJC is a bi-national organization created in 1909 to consider water and related issues along the U.S.-Canada boundary. In 1999, the IJC prepared a plan of study on the effects of water level and flow regulation on various stakeholder interests in the basin, and it established the Lake Ontario-St. Lawrence River Study Board to implement that plan.

As in its own mandate, the IJC directed the Study Board and its committees to, "assure that all significant issues are adequately addressed," serve in "their personal and professional capacities, and not as representatives of their countries, agencies, organizations, or other affiliations," and "endeavour to conduct all of their work by consensus" (IJC, 2000, pp. 2-3). This charge extended to the IJC's nearly century-long deliberations on water supply, navigation, and hydropower into the broader domains of environmental, coastal, recreational, and participatory dimensions of Great Lakes water regulation.

The Study Board adopted a "shared vision" planning approach to its five-year program of research on the effects of water level and flow regulation. Shared vision planning involves a collaborative process of water resources inquiry, systems modeling, and stakeholder participation that strives to converge on water regulation plans worthy of consideration by the IJC. The study board commissioned scores of studies involving over 150 specialists on topics that included wetlands, species at risk, coastal erosion, and flooding, selected summaries of which are reviewed in this report.

Toward the end of the five-year study period, the IJC arranged with the U.S. National Research Council (NRC) and Royal Society of Canada (RSC) to carry out this independent review of studies, reports, and models prepared for the study board, including its shared vision model.

A special committee was formed for the assignment and to prepare this report. The committee worked intensively and on a fast-track between June and August 2005 and held two meetings. The first meeting was on June 13-15, 2005 at Niagara-on-the-Lake, Ontario, Canada in a workshop setting, which allowed the committee to extensively interact with IJC and the Lake Ontario-St Lawrence River Study scientists. The workshop setting also provided a venue for the committee to receive explanations on study questions and points of clarification. After various experts (study leads) made presentations, the committee made exhaustive efforts to gain clarity on review topics.

The committee wishes to thank the following presenters: Joseph Atkinson, University at Buffalo; Lisa Bourget, IJC; Joe De Pinto, LimnoTech, Inc. ; Jana Lantry, New York State Department of Environmental Conservation; Wendy Leger, Environment Canada; Todd Redder, LimnoTech, Inc.; Albert Schiavone, New York State Department of Environmental Conservation; Eugene Stakhiv,

IWR and Study Board; Andrè Talbot, Environment Canada; William Werick, IJC Study Team; Douglas Wilcox, USGS, Great Lakes Science Center, and Peter Zuzek, Baird & Associates. The committee also wishes to thank the following for participating in this meeting: Tom McAuley, IJC; Mark Colosimo, IJC; Anthony Eberhardt, IJC-Buffalo; Ted Hullar, ISLRBC-US; Mike Shantz, Environment Canada; and Russ Trowbridge, IJC.

A second and final meeting, held on July 13-15, 2005 in Washington, DC, provided another opportunity to fill information gaps, especially in technical documentation. Also, the committee formulated and deliberated the report's major recommendations, and made plans on how to complete the report. The committee wishes to thank the following for participating in the panel discussion on flood erosion and prediction system: Guy Meadows, University of Michigan; Keith Bedford, Ohio State University; and David Schwab, NOAA, Great Lakes Environmental Laboratory.

I want to thank committee members for their dedication to reaching consensus and hard work in preparing this review. I especially thank Dr. Lauren Alexander, Study Director for the NRC Water Science and Technology Board, who organized the overall effort from beginning to end and who as a wetland ecologist contributed to all sections of the report. Ellen de Guzman efficiently kept the committee apprised of documents, deadlines, and logistics. Stephen Parker, Director of the Water Science and Technology Board, lent his broad experience to the project.

Finally, we thank the Study Board for candid presentations and discussions of Shared Vision Modeling and its associated scientific challenges. We hope the International Joint Commission finds this review useful in its deliberations.

This report has been reviewed in draft form by individuals chosen for their diverse perspectives and technical expertise, in accordance with procedures approved by the NRC's Report Review Committee. The purpose of this independent review is to provide candid and critical comments that will assist the institution in making its published report as sound as possible and to ensure that the report meets institutional standards for objectivity, evidence, and responsiveness to the study charge. The review comments and draft manuscript remain confidential to protect the integrity of the deliberative process. We wish to thank the following individuals for their review of this report:

Keith Bedford, Ohio State University, Columbus
Patrick L. Brezonik, University of Minnesota, Minneapolis
Joan G. Ehrenfeld, Rutgers University, New Brunswick
David Green, McGill University, Montreal, Quebec
Jeffrey A. Hutchings, Dalhousie University, Halifax, Canada
David Schwab, NOAA/Great Lakes Environmental Research Laboratory
Kurt Stephenson, Virginia Polytechnic and State University, Blacksburg
André St. Hilaire, Université du Québec, Canada

Although the reviewers listed above have provided many constructive comments and suggestions, they were not asked to endorse the conclusions and recommendations nor did they see the final draft of the report before its release. The review of this report was overseen by George Hornberger, University of Virginia, appointed by the National Research Council, he was responsible for making certain that an independent examination of this report was carried out in accordance with institutional procedures and that all review comments were carefully considered. Responsibility for the final content of this report rests entirely with the authoring committee and institution.

James L. Wescoat, Jr., Chair
University of Illinois at Urbana-Champaign

Contents

Summary

INTRODUCTION AND STATEMENT OF TASK

Water regulation has a complex history in the Lake Ontario-St. Lawrence River (LOSLR) basin that dates back to the mid-20th Century. In 1956, the International Joint Commission (IJC) of Canada and the United States adopted a water regulation plan for the Moses-Saunders power plant and dam, which affects water levels and flows in Lake Ontario and the St. Lawrence River (LOSLR). This initial plan required a series of adjustments in the early 1960s that culminated in 1963 with Plan 1958D. Due to the dynamics of the LOSLR system, Plan 1958D has required further deviations over the past 42 years, leading to the current situation described as Plan 1958DD (i.e., 1958 D with Deviations). Changing water management demands in the LOSLR system have made this plan outdated in many ways, and the IJC is seeking a suitable plan to replace it. In the 1990s, the IJC determined that the replacement plan will need to serve a fuller range of uses, including environmental, coastal, and recreational interests along with traditional navigation, hydropower, and municipal uses. New considerations have entered into the decision making process to select a plan: sound scientific foundations, public participation, transparency in plan development and evaluation, and inclusion of environmental considerations.

These aspects have become the hallmarks of the *Lake Ontario-St. Lawrence River Study* (LOSLR Study or "The Study"). The LOSLR Study is a 5-year, more than $20 million effort commissioned by the IJC to formulate, evaluate, and provide bases to select water regulation plans to replace Plan 1958DD. The LOSLR Study was designed to address a broad range of economic interests and environmental values affected by water level-fluctuations. The IJC Study Board, a bi-national committee charged with overseeing the LOSLR Study, commissioned empirical and modeling studies of wetlands, species at risk (SAR), recreational boating, fisheries, coastal erosion and flooding, commercial navigation, hydropower, industrial, municipal and domestic water intakes, public information and education, and hydrologic modeling. It used a Shared Vision Modeling (SVM) approach to compile, analyze, and display the results from these empirical studies; hydrologic, coastal, economic, and environmental models; and stakeholder input.

Near the middle of the final year of the 5-year LOSLR Study, the IJC asked the U.S. National Research Council (NRC) and Royal Society of Canada (RSC) to provide an independent scientific review of selected LOSLR studies, reports, and models. The NRC and RSC agreed to review the LOSLR Study materials in terms of their appropriateness and sufficiency to inform decisions related to regulation plan options. The IJC presented the committee with documents on the Shared

Vision Model (SVM), the Flooding and Erosion Prediction System (FEPS), Wetlands, Species at risk, and an Integrated Ecological Response Model (IERM).

This study was a fast-track effort: the IJC selected the documents for review and provided them to the NRC in May 2005; the NRC/RSC review appointed a special, bi-national committee of experts to carry out the assignment; the committee met once in Canada and once in the United States in June and July 2005, respectively; and the report was completed in October 2005.

The agreed upon Statement of Task (Box S-1) charged the committee to investigate (A) whether the studies and models employ reasonable scientific methods,

BOX S-1
Statement of Task

The Committee shall perform an independent review of the Lake Ontario/St. Lawrence River reports in the following areas: wetlands science and species–at–risk, the Flood Erosion and Prediction System (FEPS), the Integrated Ecological Response Models (IERM), and the Shared Vision Model (SVM). The level of emphasis for these various areas shall be approximately as follows: wetlands 15%, species-at-risk 15%, FEPS 10%, IERM 20%, SVM 40%, and reflect the International Joint Commission's (IJC) determination of its priorities in this effort.

The overarching charge shall be to evaluate the appropriateness and sufficiency of the studies and models used to inform decisions related to regulation plan options. Recommendations shall be limited to those deriving from this overarching charge and shall not address management or policy issues.

The Lake Ontario/St. Lawrence River program science, as represented in the reports and model documentation provided, will be reviewed by in terms of the degree to which:

(A) the studies reflect reasonable scientific methods, assumptions and supported findings;

(B) the models sufficiently and appropriately integrate and display the key information needed for a comprehensive evaluation and understanding of the tradeoffs for selecting among the candidate regulation plans; and

(C) the models and reports are sufficient and appropriate to evaluate the various candidate regulation plans and impacts of changes in water levels and flows.

The review shall be limited to critical evaluation and decision components of the topics listed that relate directly to the Lake Ontario/St. Lawrence River regulation plan options. This requirement shall further be interpreted to restrict the review to the impact of changing regulation levels and flows, within the limits that these two factors can be managed using the currently existing control structures and the hydrology/hydraulic characteristics of the system. The review shall neither compare regulation plan options nor provide advice on the preference of one regulation plan option over another, as these actions fall directly within the decision-making responsibilities of the Commission.

assumptions, and supporting findings; (B) how well the models integrate and display information needed for a comprehensive evaluation and understanding of tradeoffs among regulation plans; and (C) whether the models and reports are appropriate and sufficient for evaluating regulation plans and the effects of water level changes. In fulfilling the charge, the committee evaluated the selected review documents for their appropriateness and sufficiency.

The committee developed ten criteria to evaluate the appropriateness and sufficiency of the LOSLR documents presented for review in Charges A, B, and C. These criteria reflect common expectations for scientific and technical work. The ten criteria were used to evaluate IJC documents with respect to the three charges of the statement of task (Box S-2).

SCOPE AND LIMITS OF THIS REVIEW

There are five qualifications to keep in mind when using this review of the LOSLR Study.

- **Documents presented for review.** The documents presented for review were in various stages of completion. In cases where documents were incomplete, the committee tried to procure the most current version of the work. The committee treated the documents presented by the IJC as representative of the science under review, recognizing that some documents would be modified after the review.

BOX S-2
Ten Evaluation Criteria used to Review Appropriateness and Sufficiency of LOSLR Studies and Models

Charge A: Scientific Foundations
1. Empirical Foundations (e.g., sampling, analysis)
2. Quality Assurance and Quality Control (e.g., model validation, verification, and calibration; use of expert judgment; and independent peer review)
3. Treatment of Error and Uncertainty

Charge B: Integration and Display of Key Information
4. Linkages and Feedback Among Related Studies and Models
5. Spatial and Temporal Resolution and Scaling
6. Thorough Documentation
7. Effective Scientific Communication

Charge C: Overall Appropriateness and Sufficiency
8. Breadth of Study Scope
9. Balance between Scientific and Practical Professional Approaches
10. Identification of Future Study Needs

- **Selected LOSLR studies and models are reviewed.** The committee reviewed selected documents for the SVM, FEPS, and environmental sciences that represent a fraction of the body of LOSLR research undertaken. The selection of material presented for NRC/RSC review may not be fully representative of the LOSLR research and study effort as a whole. The scope of review does not encompass all of the scientific fields in the LOSLR study (e.g., hydrology and hydraulics, navigation, hydropower, municipal and industrial, river shore protection model, and other study aspects not reviewed). To the extent that the Shared Vision Model incorporates results from these other fields, this review offers a partial perspective on the overall sufficiency of LOSLR studies and models.
- **The review occurs toward the end of the 5-year LOSLR Study.** The NRC review was initiated in the final year of the LOSLR Study. Some of the draft documents presented for NRC review were being completed concurrently with this review. This timing allows LOSLR authors a chance to identify opportunities to refine drafts prior to completion, although some recommendations would have been more useful at the beginning or middle of the study period rather than this close to the Study's completion.
- **The review concentrates on the science for evaluating water level and flow effects of RPOs and for informing decision makers, and not on the RPOs themselves or on decision making policies.** Ten scientific evaluation criteria were used to evaluate the LOSLR studies and models. These criteria are common to the scientific and practical professional disciplines involved in evaluating complex studies, such as the water level and flow effects of regulation plan options in the Lake Ontario-St. Lawrence River system.
- **The review distinguishes among conclusions and recommendations in terms of their certainty, importance, and ability to fix deficiencies.** The conclusions of this report vary in terms of their certainty due to the state-of-the-science in different fields and gaps in study documentation. Some conclusions have more importance to the success of the LOSLR Study than others. Points of study weaknesses and recommendations vary in the degree to which they can be fixed and the amount of time and additional research needed to address them. The review strives to distinguish among recommendations that entail short- and long-term action. Short-term recommendations are largely limited to improving the documentation, scientific communication, and disclosure of potential implications of these limitations for decision makers. Longer-term recommendations require investment in additional data collection, analysis, and interpretation.

KEY CONCLUSIONS AND RECOMMENDATIONS

The LOSLR Study's breadth is impressive, and the scale and inclusiveness of the studies and models are commended. In terms of informing decision making, however, the reviewed studies and models show deficiencies when evaluated against the ten criteria. Four overarching conclusions are drawn from the review of these documents, and one additional prospective recommendation is made to build upon the significant body of work already completed in the LOSLR Study.

1. LOSLR studies and models expand interdisciplinary scientific inquiry on the potential environmental effects of water level and flow regulation options in the Lake Ontario-St. Lawrence River Basin in ways that are useful for informing some aspects of decision making. The LOSLR studies undertook a broad set of studies and developed models that extend beyond previous Great Lakes water regulation efforts in compiling results of scientific analysis and stakeholder input, and they result in some notable successes. First, given the complexity of the LOSLR system, bi-national interests, and the range of scientific and other information compiled, the undertaking of this comprehensive study is a major contribution by itself. Further, identification and inclusion of environmental performance indicators advance understanding of the LOSLR system and water resources planning. The LOSLR studies also have created large new databases on wetland vegetation and coastal land use, that did not exist previously and that could, if archived and made readily accessible, have continuing value. This organization of a multi-disciplinary water resources planning project could serve as an example for other regions concerned with water level and flow regulation.

With few precedents for a study of this scale on regional water level regulation, opportunities for improvement are expected in the LOSLR effort. The following three findings and associated recommendations indicate the need for scientific and technical improvements in relation to the three charges in the statement of task.

2. The scientific foundations of the studies and models presented for review vary widely in empirical support, and overall, need stronger and more consistent quality control, quality assurance, and treatment of error and uncertainty to inform decision making. Three evaluation criteria were used to assess the scientific foundations of the LOSLR studies and models presented for review: empirical support, quality assurance, and treatment of error and uncertainty.

- **Empirical support (e.g., data, sampling, analysis).** In the LOSLR Study documents reviewed (wetlands, species at risk, and IERM), empirical research was conducted in coastal and environmental investigations, and some problems were noted. In the coastal research (FEPS model and sub-models), a detailed land use parcel database was developed, but that database differs in completeness for Canada (~75 percent coverage) and the USA (~100 percent coverage), but neither the means to complete the Canadian database nor actions to account for differences in data coverage were included in the documentation.

The environmental work depends on wetland and species at risk empirical data. The wetlands studies provided detailed accounts of empirical sampling, which allowed for detailed evaluation of this work. However, wetland sampling appears to have been limited primarily to shallow water sites; it excluded or undersampled deeper-water wetlands, which may have resulted in an underestimation of high quality habitat associated with deeper water wetland ecosystems. Also unclear is the degree to which the sampled wetlands are representative of wetland vegetation types across the LOSLR shoreline. The reviewed documents do not present evidence that wetlands were selected randomly. Furthermore, quantitative methods were not documented to show how findings in the sub-set of wetlands that were sampled can be extrapolated to LOSLR wetlands in general. In the SAR work, SAR performance indicators were developed inconsistently: some were developed using empirical data, but others were developed via different means (e.g., expert judgment).

The use and application of regression models in the SVM and FEPS work needed better explanation and defense. In the SVM research, regression equations rather than hydrodynamic models were used to calculate water levels and flows when the latter approach presumably would have been empirically feasible and more accurate. In the FEPS research, regression analysis problems concerned the variable quality of the empirical data used in the regression analyses and the 4th and 5th order polynomials that may be numerically unstable and misrepresent the potential effects of extreme events. The rationale for using regression and limitations of the regression analyses were not fully discussed or quantified in the documents reviewed.

Recommendation: As no new data can be collected in the near-term, LOSLR study final reports should identify limitations of empirical data and information sources, data gaps, and sampling problems, and discuss their implications for decision making. For the longer-term, research to correct data and model deficiencies, including replacement of regression equations with process models, should be prioritized.

- **Quality assurance and quality control (e.g., model validation, verification, and calibration; use of expert judgment; and independent peer review).** In general, reviewed models lack adequate validation, verification, and calibration. In some cases, validation may have occurred or is briefly mentioned but is not thoroughly documented; in others, it appears not to have been undertaken. For example, the SVM is still awaiting model validation from technical work group members. Reports on FEPS suggest that model calibration has occurred, but that documentation is not included in the materials presented for review. Documentation of the proprietary COSMOS model (a sub-model of FEPS) referred to, yet did not provide, model validation; even though proprietary models too, need to be validated and subjected to full scientific peer review.

Environmental information presented for review lacks demonstrated protocols for quality assurance, and the IERM model acknowledges that validation was not attempted. The SVM and FEPS models and SAR studies make creative use of expert judgment, but such judgment should be subject to formal quality assurance measures and standard methods for eliciting expert judgments. In cases where peer review was documented, it was inconsistent. Some studies used "peer review" by fellow team members while others involve refereed papers. This NRC/RSC report is the only known independent scientific review of the broader LOSLR study program, and a review earlier in the Study's five-year lifespan would have been more timely for identifying and rectifying deficiencies.

Recommendation: In the short-term, LOSLR final reports should inform decision makers of the types of quality assurance measures that were and were not undertaken and discuss their potential implications for decision making. Further independent scientific review of final reports is recommended. In the longer-term, rigorous quality assurance methods should be put in place for evaluating the effects of water level and flow regulation.

- **Treatment of error, uncertainty, and risk.** The treatment of error, uncertainty, and risk in the studies and models reviewed was neither commensurate with scientific and practical standards nor conducted at a level suitable to inform decision making. The SVM, FEPS, and IERM models do not present an overall framework for uncertainty analysis, which should include natural variability, data uncertainties, model uncertainties, model parameter uncertainties, and decision model uncertainties. Some individual studies (e.g., wetlands vegetation analysis) address natural variability and indicate error bars. The SAR 3A report provides a good model for qualitative discussion of uncertainty. In contrast, the SVM treats the uncertainty of environmental performance indicators with a simplistic and unexplained 10 percent criterion, and it does not apply any uncertainty estimate to economic indicators. Linkages among LOSLR studies and models inherently propagate uncertainties, but SVM documentation does not analyze those cumulative uncertainties or discuss their implications for informing decision making. Without formal analysis and discussion, it is not possible to assess the types or magnitudes of error and uncertainty for particular water regulation plans, or to know whether differences between plans are significant.

Recommendation: In the short-term, LOSLR final study reports should inform decision makers of the uncertainties that were analyzed, those that were not analyzed, and their potential implications for decision making. Future studies of water level regulation effects in the LOSLR basin should develop a comprehensive approach to uncertainty analysis.

3. The LOSLR models and studies reviewed here do not adequately integrate and display the key information needed for comprehensive evaluation and understanding of the tradeoffs among the candidate regulation plans. This conclusion is based on the following four review criteria:

- **Linkages and feedbacks among related studies and models.** "Comprehensive evaluation and understanding of tradeoffs among regulation plan options alternatives" (NRC Committee Statement of Task Charge B, Appendix A) requires a system dynamics approach that models the linkages and feedbacks among socioeconomic and environmental processes. The SVM compiles first-order effects on environmental, coastal, and other indicators generated by FEPS, IERM, and other models. But, as the IERM user's manual indicates, it is not an ecosystem model that incorporates the feedback effects of water level variation on species and habitat conditions. Instead, it compiles initial impacts (first-order effects) on performance indicators, and it is thus an impact accounting model rather than an ecosystem model. In terms of model linkages, the FEPS model alters the bathymetry of shoreline environments, but those bathymetric changes were not fed into the IERM to vary wetland inundation, which could be used to model vegetation, shoreline habitats, and other environmental performance indicators associated with water level variation. These vegetation changes could have feedback effects on sediment transport and coastal erosion. External model linkages include economic and demographic scenarios that are relevant for evaluating candidate water regulation plans to replace Plan 1958DD. For example, real estate values of coastal property continue to rise at rapid rates, and the demand for different water and related land uses is changing, but the SVM does not incorporate such scenarios in its structure.

This report acknowledges that some of these linkages and feedbacks require knowledge beyond the current limits, and that fact should be discussed in the final reports and presentation of SVM results as well. Other linkages and feedbacks between the SVM and its sub-models, and externally between the SVM and scenarios of socioeconomic change, could have been addressed. The reviewed studies and models make progress toward comparing the effects of regulation plan options, but the comparisons do not provide a comprehensive basis for evaluating and understanding tradeoffs among regulation plan options.

Recommendation: In the short-term, the LOSLR final reports should inform decision makers of what has, and has not yet, been accomplished in the way of integrated water and environmental systems modeling. As part of an ongoing program, a LOSLR modeling system that dynamically links and reflects feedback among sub-models is recommended.

- **Treatment of spatial and temporal resolution and scaling.** Scaling issues in the Lake Ontario-St. Lawrence River basin are challenging. The LOSLR studies involve a wide range of spatial and temporal scales, and associated concerns. For example, although more detailed hydrologic time series and station data are available at multiple locations on Lake Ontario and at a finer time step than the quarter-month period, the STELLA model in the SVM generates a single series of quarter-monthly values for the level of Lake Ontario, based on historical water management practice. Use of these single series values can result in a loss of precision, as the quarter-month does not provide enough temporal variation for many environmental impacts, including fish, SAR, and wetlands. This coarse time step was recognized as a potential problem in the LOSLR *Plan of Study*, which called for a 2D hydrodynamic model for the St. Lawrence River that operated on fine enough time scales to supplement the quarter-monthly time step generated by the SVM. As noted earlier, the LOSLR approach of using quarter-monthly values in Lake Ontario to calculate water levels for selected stations in the upper St. Lawrence River through regression analysis is inferior to hydrodynamic flow routing, and the combined use of regression and hydrodynamic models in the LOSLR Study needs to be more fully explained. The FEPS model uses lake level elevations along with a grid of wind and wave fields that erode and flood individual shoreline parcels and reaches, the results of which are then aggregated back to lake-wide effects. The errors and uncertainties associated with these different resolutions and scales of inquiry need fuller analysis and discussion, as errors may exceed the differences among model outputs for some performance indicators and plans.

Recommendation: In the short-term, the LOSLR final reports should inform decision makers of temporal and spatial scaling issues that affect the accuracy and uncertainty of predictions of regulation effects. In the longer term, choice of time step should better reflect the critical response times for system indicators, including those where transient fluctuations in water temperature and water level are critical, and appropriate hydraulic and hydrodynamic modeling approaches should be implemented.

- **Documentation of scientific studies and models.** Of the ten criteria employed in this review, inadequate documentation is the most apparent deficiency, with examples throughout the materials presented for review. Fortunately, FEPS included more detailed descriptions of modeled performance indicators than other studies, but did not document the models themselves. Descriptions of wetland methodologies need additional information about site selection and means to ensure adequate representativeness of sampled sites. A user's manual exists for the IERM and provides partial documentation, but explanations of weighting and aggregation in the model are insufficient. Exceptions to these general patterns include the SAR 3A and 3B reports, which are well documented. Better documentation is needed to explain choices of what was done and methods used, and the rationale behind those decisions. The SVM is the primary tool for understanding and evaluating tradeoffs among potential regulation plans. It was surprising, therefore, that the SVM had the least amount of documentation presented for this review, and the documentation that was presented was not at a level of completion ready for external scientific review. Documentation of the SVM should have a more complete discussion of its role in the Shared Vision planning process; describe SVM development and refinement, including standard technical documentation of all component models; and describe how scientific and stakeholder criteria were used interactively to formulate, screen, and evaluate the range of choice among regulation plan options.

Recommendation: In the short-term, LOSLR final reports should include a thorough documentation of studies and models, especially the Shared Vision Model, and seek further independent scientific review of those reports.

- **Effective scientific communication.** Effective scientific communication is achieved when scientific information is presented to and received and correctly understood by scientific, public, and decision making groups. The efficacy of scientific communication varies among LOSLR studies and models in the materials submitted for review. Performance Indicators, an Index of Ecological Integrity, and documentation of studies, models, and sub-models were used to communicate scientific information from the LOSLR Study. In general, the environmental studies and performance indicator summaries were easier to understand than the sub-models' documentation, and sub-model documents were more digestible than the

SVM documentation. An example of deficient, or even misleading, communication is the differential treatment of economic and environmental indicators in which the former are presented as simple values while the latter are subject to a +/-10 percent error. The LOSLR Study's display of model output in a spreadsheet file of tables and graphs, known as the "Board Room," has strong potential as a venue for scientific communication.

Recommendation: In the short-term, the LOSLR final reports should communicate their scientific results with transparency to support decision making while giving a full treatment of uncertainties and non-scientific dimensions of the studies. In the longer-term, the SVM Board Room may be refined for continuing use as a vehicle for scientific communication.

4. Despite the breadth of LOSLR studies and models, ongoing analysis is needed to provide a strong scientific basis for long-term decision making about water level and flow regulation in the Lake Ontario-St. Lawrence River basin. Three points support this conclusion. First, current knowledge about the lower Great Lakes system is not comprehensive. While the LOSLR studies and models broaden understanding about the potential effects of regulation plans, a more comprehensive data collection and modeling approach is needed to understand system feedbacks, linkages, and uncertainties. Ideally, a system dynamics model should be used to: (a) improve the physical system description; (b) identify the most important feedback relationships; and (c) improve understanding of feedback effects on system behavior. Some feedback relationships require expansion of the model boundaries so that key processes, ranging from coastal urbanization and regional economic growth to climate change, are incorporated and their impacts are made visible within the model.

Second, the LOSLR history with Plan 1958DD shows that regulation plans can be superseded by newer, better plans, and change in management objectives. Any plan adopted now on the basis of current science without provision for regular updating as knowledge advances is likely to require adjustments over time.

Third, the LOSLR models evaluate effects of *future* regulation plans and hydrologic scenarios based primarily on *historical* and *current* environmental and social performance indicators. This is important given the significance of hydroclimatic variability for water regulation and the challenges of modeling current environmental and socioeconomic processes. Although this report does not review the climate change research and scenarios, it commends the LOSLR inclusion of global processes that affect the robustness of regional regulation decisions. In the future, however, regulation plan decisions will also require comparable scenario development and evaluation for other environmental and social processes. Changes in regional economic structure, demography, water demand, transportation technology, coastal land use, and socioeconomic values will likely transform the profile of stakeholder interests, performance indicators, and socioeconomic impacts associated with water level regulation. The past half-century indicates that these types of structural shifts in socioeconomic and environmental conditions and values, in conjunction with hydrologic variability, have had substantial implications for regulation plan decision making.

The LOSLR studies and models begin to address these issues through brief conceptual narratives with a planning horizon of 10 to 15 years that are linked to the SVM. The conceptual narratives employ a common template, but they vary in detail, completeness, and level of peer review. Correcting the scientific and modeling deficiencies identified in this review is necessary and appropriate, but not sufficient, for informing water regulation decisions on a long-term multi-decadal timescale.

Recommendation: In the short-term, the LOSLR Study should complete the conceptual narratives. For the longer-term, the IJC should consider an ongoing management and monitoring system to feed the results of current choices for water level regulation into a dynamic model of the LOSLR system to strengthen the scientific basis for future planning on a multi-decadal timescale, as outlined in the final recommendation below.

Looking Ahead: Adaptive Management in the LOSLR Basin

As the LOSLR Study draws to a close in 2005, a unique opportunity is presented for water level regulation in the LOSLR basin. Even after the deficiencies noted above are addressed, and a new regulation plan is adopted and implemented, the need will remain to monitor the system for responses to the new regulation plan. Long-term monitoring may also indicate needed adjustments to the plan. Adaptability is mentioned in the LOSLR *Vision, Goals, and Guiding Principles* in a number of ways: "... regulation plans will incorporate flexible management...;" "Regulation of the Lake Ontario-St. Lawrence River System will be adaptable...;" and "...regulation plans will incorporate...flexibility to adapt...." An adaptive management program could help the basin constituents build upon the LOSLR studies and models over time.

Before an adaptive management program is designed, the deficiencies noted in LOSLR models and studies need to be corrected to avoid perpetuating existing problems. The challenges of implementing an adaptive management in the Lake Ontario-St. Lawrence River basin should not be underestimated. Adaptive management can be resource intensive: an "active" adaptive management plan could involve annual costs comparable to those of the LOSLR study; "passive" adaptive management costs would be significantly lower, depending upon the scope of monitoring and management involved, but also less useful. Either way, adaptive management is seen as a viable option to build upon the LOSLR Study successes, address deficiencies, and maintain a responsive, flexible water regulation plan for the LOSLR basin.

Recommendation: In the short-term, adaptive management alternatives should be identified that build upon the LOSLR studies and models. In the longer term, the IJC should, in collaboration with other scientific and stakeholder organizations in the basin, develop an adaptive management program that would provide a continuing scientific basis for monitoring the effects of water regulation, experimenting with alternatives, and thereby improving decisions about future regulation plan options.

Chapter 1
Introduction

In 1999, the International Joint Commission (IJC) commissioned the Lake Ontario-St. Lawrence River (LOSLR) Study as a 5-year, more than $20 million study. In identifying scientifically-based adjustments to water level and flow regulation, the LOSLR Study's purpose is to (1) review the current regulations of levels and flows in the LOSLR system with consideration for the impact of regulation on a variety of interests; (2) build a progressive understanding of the system; and (3) provide a strong scientific foundation for deciding future regulation plan options. At the broadest level, the vision of the LOSLR Study[1] is, "to contribute to economic, environmental and social sustainability of the Lake Ontario and St. Lawrence River System." The goals of the Study are to identify flow regulation plans and criteria that serve the range of stakeholder interests, are widely accepted, and take into account hydroclimatic conditions in the basin.

The LOSLR Study was designed within the context of seven guiding principles[2]:

1. Criteria and Regulation Plans will contribute to the ecological integrity of the Lake Ontario-St. Lawrence River System ecosystem.
2. Criteria and Regulation Plans will produce a net benefit to the Lake Ontario-St. Lawrence River System and its users and will not result in disproportionate loss to any particular interest or geographic area.
3. Criteria and Regulation Plans will be able to respond to unusual or unexpected conditions affecting the Lake Ontario-St. Lawrence River System.
4. Mitigation alternatives may be identified to limit damages when considered to be appropriate.
5. Regulation of the Lake Ontario-St. Lawrence River System will be adaptable to the extent possible, to accommodate the potential for changes in water supply as a result of climate change and variability.
6. Decision making with respect to the development of the Lake Ontario-St. Lawrence River System Criteria and Plans will be transparent, involving and considering the full range of interests affected by any decisions with broad stakeholder and public input.
7. Criteria and Regulation Plans will incorporate current knowledge: state-of-the-art technology; and the flexibility to adapt to future advances in knowledge, science, and technology.

---

[1] Full text of the vision is available online at *http://www.losl.org/about/vision-e.html*; August 2, 2005.
[2] Available online at *http://www.losl.org/about/vision-e.html*.

These economic, environmental, and social criteria represent an unprecedented expansion of inquiry related to water regulation in the basin. Although not exhaustive (as discussed later in this report), the LOSLR scope extends far beyond the historical preference given to domestic water use, navigation, hydropower, and irrigation (Boundary Waters Treaty, 1909, Article VIII).

This chapter introduces the Lake Ontario St. Lawrence River Study (LOSLR Study or "the Study") and the independent review of it by the National Research Council (NRC) in collaboration with the Royal Society of Canada (RSC). The first part of the chapter introduces the Lake Ontario-St. Lawrence River system, including its geographic setting and the water level and flow regulation issues that prompted the LOSLR Study. Next is a brief outline of the LOSLR Study, including its scope and organization. The final section of the chapter describes the NRC independent review, the statement of task from the IJC, the evaluation criteria used to address the statement of task, and the organization of chapters that follow.

LAKE ONTARIO-ST. LAWRENCE RIVER BASIN

Geographic Setting

Shared by the United States and Canada, the Lake Ontario-St. Lawrence River (LOSLR) basin drains the largest freshwater lake system in the world (Figure 1-1). It includes the lower Niagara River, Lake Ontario, and the St. Lawrence River basin. The lower Great Lakes shared international waters extend from the lower Niagara River, downstream from Niagara Falls, through Lake Ontario and the upper St. Lawrence River to just downstream from the Moses-Saunders Dam near the towns of Cornwall, Ontario and St. Regis and Massena, New York. (Figure 1-1). From there, the St. Lawrence River flows northeast through the Province of Quebec, Canada, until it finally discharges into the Gulf of St. Lawrence.

The LOSLR basin supplies drinking water for some 8.6 million people. It supports complex aquatic, wetland, and coastal ecosystems affected by water level and flow fluctuations. It serves as a navigation route for global and regional maritime shipping from the Port of Montreal through the St. Lawrence Seaway that depends upon reliable flow regulation. Hydropower production at the Moses-Saunders facilities currently averages 13 million megawatt hours per year. Recreational boaters and waterfront communities enjoy the beautiful water body and coastal environments, though shoreline properties, municipalities, and infrastructure face flooding and erosion hazards.

Lake Ontario drains 64,030 km^2. It has the smallest surface area of all of the Great Lakes (18,960 km^2), but the second greatest average depth (86 m). Lake-wide monthly water levels over the past century range from an historic low of 73.74 m (December 1934) to a record high of 75.76 m (June 1952) (USACE, Detroit, 2005; IGLD 1985). A difference of even a half-meter in water levels can aggravate flooding, erosion, boating problems, wetland habitat and fish spawning conditions.

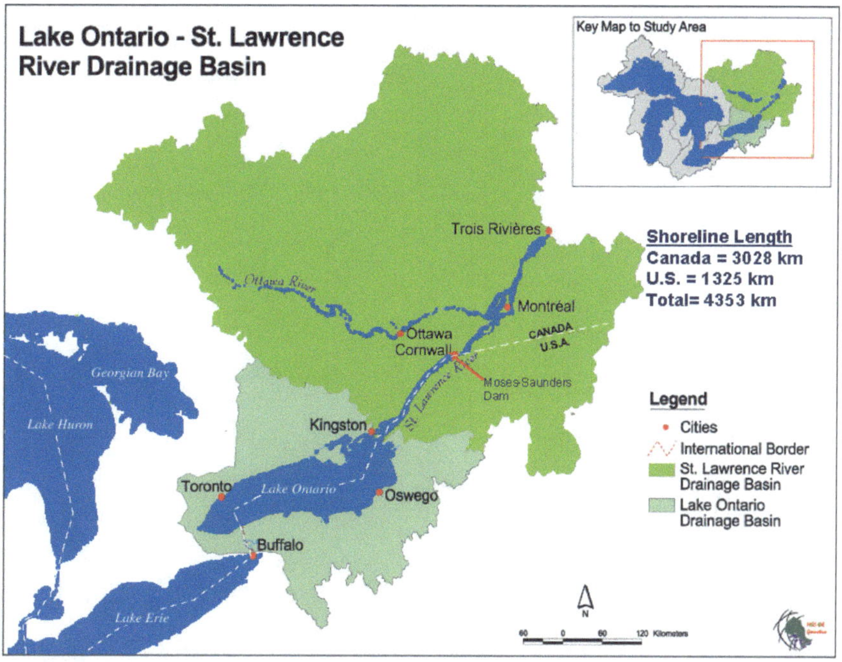

FIGURE 1-1 Map of LOSLR drainage basin. SOURCE: Map courtesy of the IJC.

Under the Boundary Waters Treaty of 1909, the Canadian and United States governments established the International Joint Commission (IJC) to address mutual water-related concerns (Boundary Waters Treaty, 1909; available on-line at *http://www.ijc.org/rel/agree/water.html*), such as water level and flow regulation and their effects.

Water Regulation in the Lake Ontario-St. Lawrence River Basin

Water regulation activities in the LOSLR system date to the mid-20th century. Figure 1-2 describes the water level regulation institutional organization for the Lake Ontario-St. Lawrence River. The 1952 Order of Approval, amended in 1956, authorized construction of the Moses-Saunders Dam, Iroquois Dam, and Long Sault Dam in the International Rapids Section of the St. Lawrence River (Figure 1-3). The 1952 Order of Approval also created the International St. Lawrence River Board of Control (ISLRBC or Board of Control) to operate the Moses-Saunders Dam and power plant, completed in 1958, which are the principal structures affecting water levels and flows (Figure 1-4).

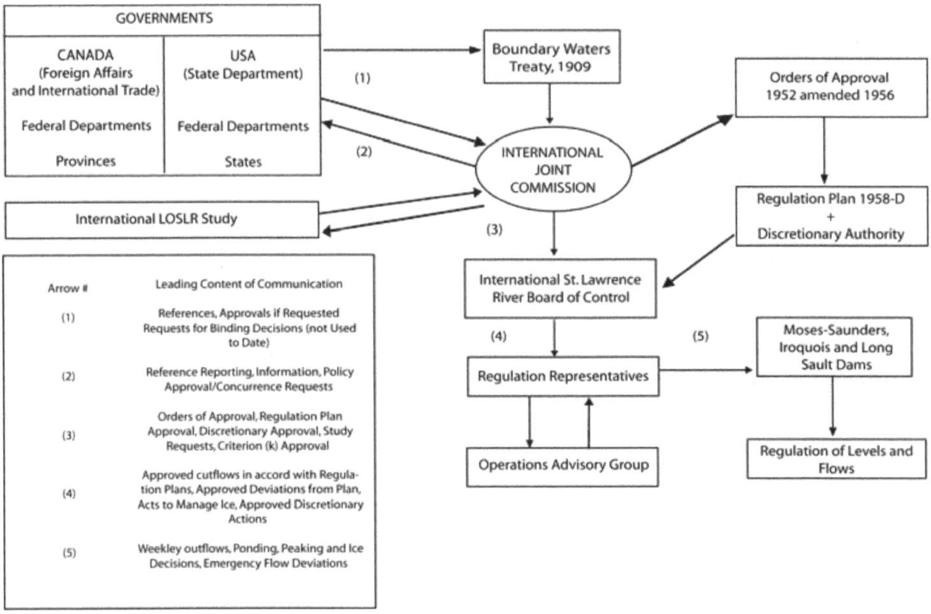

FIGURE 1-2 Levels Regulation Institutional Structure. SOURCE: Clinton Edmonds and Associates (2002, p. 15).

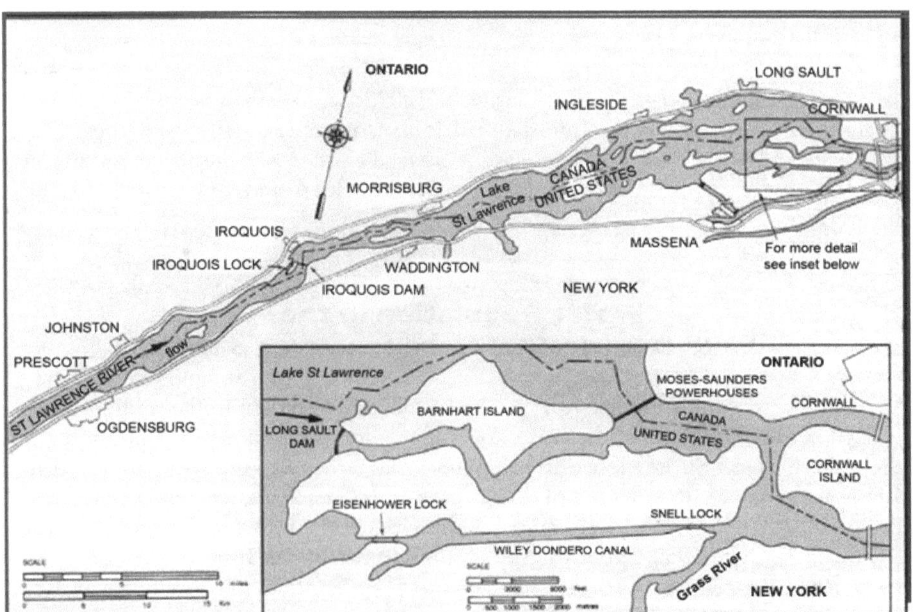

FIGURE 1-3 Location of Upper St. Lawrence River Control Structures. SOURCE: ISLRBC (2004, p. vii).

FIGURE 1-4 Moses-Saunders Dam. Photo courtesy of James Wescoat.

In the early 1950s, wide fluctuations in water levels led to a series of changes in the regulation plan, culminating in Plan 1958D, adopted in 1963 (Clinton Edmonds and Associates, 2002). Figure 1-5 shows average water levels from 1918 to 2003, which suggests that 1958D has reduced extreme high and low lake levels (IGLD, 1985), and increased mean water levels.

The St. Lawrence Board of Control is authorized to deviate from Plan 1958D in response to emergencies and within reasonable limits to reduce negative impacts and increase net benefits for stakeholders. By the late 1990s, deviations reportedly occurred over half of the year, leading to a *de facto* regulation plan known as Plan 1958DD (Plan 1958D with Deviations). The extent of deviations from Plan 1958D has raised the question of whether another regulation plan could provide greater benefits and fewer losses than Plan 1958D.

As the IJC considers possible attributes of a new water regulation plan, it will use the LOSLR study to provide scientific and practical bases for deliberating among candidate regulation plan options. The IJC has been clear about certain components needed to improve upon 1958D for regulating water levels in the LOSLR basin: sound scientific foundation, public participation, transparency in the development and decision making process of regulation plans, and environmental considerations will be included in the 1958D regulation plan successor.

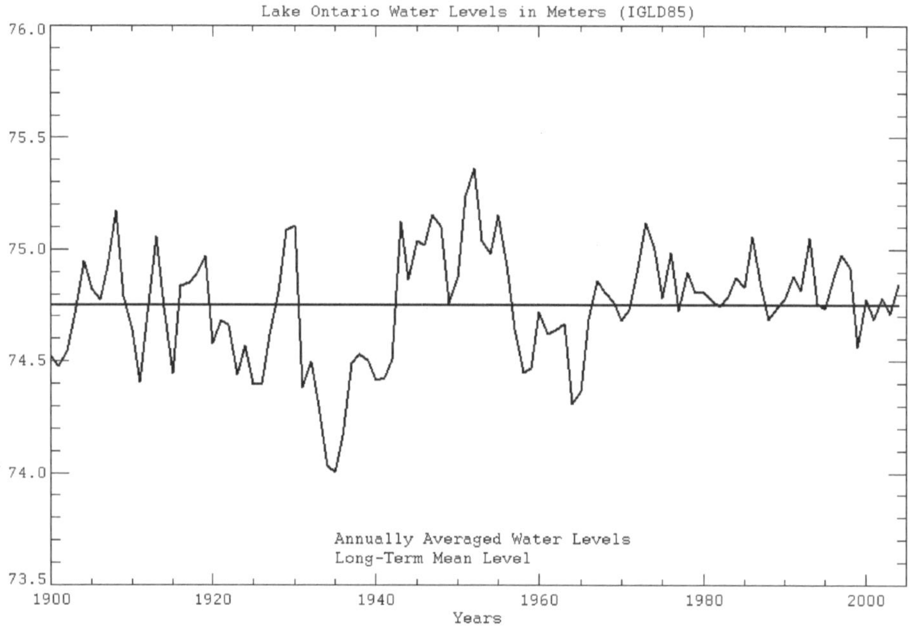

FIGURE 1-5 Lake Ontario Water Levels in Meters. SOURCE: IGLD85. Available online at *http://www.glerl.noaa.gov/data/now/wlevels/lowlevels/plot/Ontario.png.*

LOSLR STUDY STRUCTURE

LOSLR Study Design

The LOSLR Study was designed to address a broad range of human interests and environmental values affected by water level fluctuations. The suite of subjects in the Study reflects a broad spectrum of considerations in regional water resources planning that may serve as a case study for other regional efforts to balance several, sometimes competing, water uses. In the LOSLR Study, empirical and modeling studies were commissioned to form the scientific foundations for water regulation plans. Individual study subjects include wetlands, species at risk, recreational boating, fisheries, coastal processes of erosion and flood potential, commercial navigation, hydropower, industrial, municipal and domestic water intakes, public information and education, and hydrologic modeling. Individual studies were designed and conducted as separate investigations, and study results were integrated into the SVM. The SVM links results from the empirical studies; hydrologic, coastal, economic and environmental models; and stakeholder input (Figure 1-6). The SVM's Board Room provides a mechanism for understanding and presenting tradeoffs among the different regulation plan options.

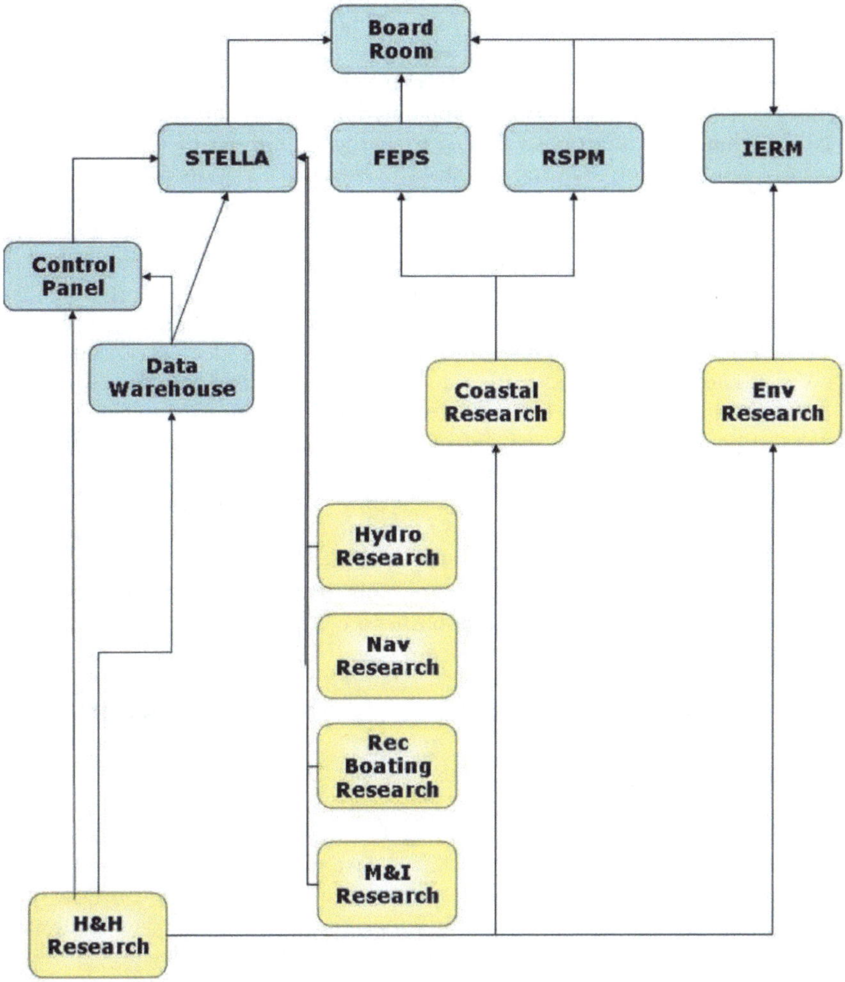

FIGURE 1-6 The Shared Vision Model structure. SOURCE: Werick (2005).

LOSLR Study Administration

Three groups have roles in the LOSLR Study effort. First, the LOSLR Study Board, which has overall responsibility for the Study, provides guidance and direction to the LOSLR study management and staff, and oversees Study scientific aspects. Fourteen people, half from the United States and half from Canada, form the International LOSLR Study Board (appointed in 2000 by the IJC).

Nine technical work groups (TWGs) form the second tier of organizations in the LOSLR Study. More than 150 scientists and technical experts from the United States and Canada work with the TWGs to carry out the scientific work of the Study by conducting empirical studies, developing models, and reviewing data. The nine TWGs are in the technical areas of: wetlands, recreational boating, fisheries and the environment, coastal processes including erosion and flood potential, commercial navigation, hydropower, industrial, municipal and domestic water intakes, public information and education, and hydrologic modeling. Results from the TWG studies are conveyed to the Study Board, integrated into the SVM, and used in the formulation and evaluation of regulation plan options.

A third group that is part of the LOSLR Study, but has no administrative, management, or oversight role, is the Public Interest Advisory Group (PIAG). PIAG was developed by the IJC to ensure healthy public participation, debate, input, and guidance related to the study process. The PIAG's role is to foster public understanding on the causes of water level problems and how possible solutions to some problems affect others. PIAG represents stakeholder interests, convenes public meetings around the basin, and publishes a newsletter, *Ripple Effects*.[3]

THE INDEPENDENT REVIEW OF THE LOSLR STUDY

Near the middle of the last year of the 5-year LOSLR Study, the IJC asked the NRC and the RSC to provide an independent review of a select body of scientific and technical work from three of the nine TWGs.[4] Given the timing of this review in the course of the LOSLR study timeline, a fast-track review in the last half of 2005 began shortly afterwards: a bi-national committee was formed in May, the first meeting was held in June, the second meeting a month later and, following external peer review, the report was submitted to IJC in October, 2005.

The Statement of Task

The overall charge to the NRC committee was to provide an independent review of the select scientific reports and models that the Study Board will use as the bases for evaluating and choosing among regulation plan options. The statement of task (Box 1-1) contains three charges to the committee (more information con-

[3] Available online at *http://www.losl.org/news/news-e.html.*

[4] The NRC provided the overall organization and management for the review.

BOX 1-1
Statement of Task

NRC Committee shall perform an independent review of the Lake Ontario/St. Lawrence River reports in the following areas: wetlands science and species at risk, the Flood Erosion and Prediction System (FEPS), the Integrated Ecological Response Models (IERM), and the Shared Vision Model (SVM). The level of emphasis for these various areas shall be approximately as follows: wetlands 15%, species at risk 15%, FEPS 10%, IERM 20%, SVM 40%, and reflect the International Joint Commission's (IJC) determination of its priorities in this effort.

The overarching charge shall be to evaluate the appropriateness and sufficiency of the studies and models used to inform decisions related to regulation plan options. Recommendations shall be limited to those deriving from this overarching charge and shall not address management or policy issues.

The Lake Ontario/St. Lawrence River program science, as represented in the reports and model documentation provided, will be reviewed by in terms of the degree to which:

(A) the studies reflect reasonable scientific methods, assumptions and supported findings;

(B) the models sufficiently and appropriately integrate and display the key information needed for a comprehensive evaluation and understanding of the tradeoffs for selecting among the candidate regulation plan (RPOs); and

(C) the models and reports are sufficient and appropriate to evaluate the various RPOs and impacts of changes in water levels and flows.

The review shall be limited to critical evaluation and decision components of the topics listed that relate directly to the Lake Ontario/St. Lawrence River regulation plan options. This requirement shall further be interpreted to restrict the review to the impact of changing regulation levels and flows, within the limits that these two factors can be managed using the currently existing control structures and the hydrology/hydraulic characteristics of the system. The review shall neither compare regulation plan options nor provide advice on the preference of one regulation plan option over another, as these actions fall directly within the decision-making responsibilities of the Commission.

cerning the statement of task, including documents reviewed can be found in Appendix A): whether the studies and models employ reasonable scientific methods, assumptions and supported findings; how well the models integrate and display information needed for a comprehensive evaluation and understanding of tradeoffs among regulation plan options; and whether the studies and models are appropriate and sufficient for evaluating candidate regulation plans and the effects of water level changes.

Two implicit aspects of the charge influenced the committee's review: the scope of the NRC review in comparison to the scope of the complete LOSLR

Study and the documents and documentation presented to the committee for review.

Scope of the Review

The statement of task clearly outlines the scientific areas that are the subject of the review. The review involves selected reports, documents, and models from three of the nine Technical Work Groups: the Environmental Technical Work Group (ETWG), the Coastal Processes technical work group (Coastal), and the Shared Vision Model (SVM) of the Plan Formulation and Evaluation work group (a list of documents reviewed is provided in Appendix A). This review represents only a small sample of the body of work from the overall LOSLR Study. Products from the hydrology and hydraulics, navigation, recreational boating, hydroelectric power, and information management work groups; the Economics Advisory Committee; and the Public Interest Advisory Group, were not included in this review, and as such, are not discussed in this report.

Given the short timeline of the review, this selectivity of material was necessary. Time was inadequate to review any more information than what was presented, but reviewing this selection of documents and models separately from products of the other six Technical Work Groups presented a contextual challenge. The NRC/RSC were asked to review the SVM—a model that integrates studies from all of the Technical Work Groups—but only had the detailed input and output of the SVM related to wetlands, SAR, the Integrated Ecological Response Model (IERM), and the Flooding and Erosion Prediction System (FEPS). Thus, the findings must be viewed as partial in scope and are presented as contingent upon a larger body of work.

Documents Provided for Review

The IJC, as per the statement of task, selected the documents that were provided to the NRC/RSC for review that pertain to wetlands, SAR, the IERM, the FEPS and the SVM (Appendix B for list of review and background materials). Weights assigned to each of these subject areas reflect the IJC's priorities and the focus of the NRC review: wetlands, 15%; species at risk, 15%; FEPS, 10%; IERM, 20%; SVM, 40%.

The IJC selected the documents for the committee to evaluate "the Lake Ontario/St. Lawrence River program science, *as represented in the reports and model documentation* ... " [italics added]. The committee interpreted this text to mean that the documents are representative of the larger body of reports and models produced in the LOSLR Study. This point is significant in light of the concurrent completion of LOSLR studies and the NRC/RSC review.

Some of the documents presented for review were in various stages of completion, or were superseded during the course of the review. In the cases where the documents were incomplete, the committee tried to procure the most recent, updated, and current version of the work. Ultimately, due to constraints of the study time line and in accordance with the Statement of Task, the committee considered the documents presented by the IJC as representative of the study science, regardless of their level of completion.

The Approach to Reviewing LOSLR Documents

The overall thrust of the NRC/RSC independent review was to determine the "appropriateness and sufficiency" of the wetlands, SAR, IERM, FEPS, and SVM work to inform decisions about regulation plan options. As "appropriateness and sufficiency" are not defined by the IJC or Study Board, the committee developed general evaluation criteria to gauge scientific appropriateness and sufficiency.

Ten evaluation criteria were selected based on scientific and practical professional approaches (Box 1-2). They include empirical foundations of scientific investigation; issues of scale, uncertainty, feedbacks, and quality assurance; and technical documentation and scientific communication. The ten criteria are related to the three specific charges within the Statement of Task (Box 1-1), and they form the basis for evaluating review documents in each of the subject areas (wetlands, SAR, IERM, FEPS, and SVM).

CHARGE A: The LOSLR program science...will be reviewed in terms of the degree to which...the studies [and models] reflect reasonable scientific methods, assumptions, and supported findings.

This charge evaluates how the scientific inquiry was conducted. It addresses the scientific questions posed, empirical methods to address those questions, and whether the findings are supported by the data. Also included in this charge is how uncertainty, error, and gaps in data were communicated; how data are managed for quality assurance and quality control; and methods to validate, verify, and calibrate models used in the LOSLR Study. In short, Charge A asks (1) whether the correct science was conducted and (2) whether the science was conducted correctly. Evaluation criteria 1, 2, and 3 (see Box 1-2) are used to address Charge A.

Criterion 1: Empirical Foundations

The empirical foundations of the studies are vital for estimating the potential effects of water level change. Strong empirical foundation includes identifying the correct thread of scientific inquiry, i.e., whether the scientific question is formulated to address the scientific problem. It is important to determine whether the methods used will enable investigators to answer the scientific inquiry. Methods include data collection, model construction, data analysis, and interpretation. Flaws at any of these steps—questions, data, methods, analyses, or interpretation—can negatively impact the scientific outcomes.

BOX 1-2
Ten Evaluation Criteria

1. Empirical foundations
2. Quality assurance and quality control
3. Treatment of error and uncertainty
4. Linkages and feedbacks among related studies and models
5. Spatial and temporal resolution and scaling
6. Thorough documentation
7. Effective scientific communication
8. Breadth of study scope
9. Balanced between scientific and practical professional approaches
10. Identification of future study needs

Criterion 2: Quality Assurance and Quality Control

Once the data have been collected, analyzed, and interpreted, some mechanism is needed to ensure that the quality of data-related actions is acceptable. Methods for QA/QC vary by discipline, but conventions exist to ensure high quality data and scientific output. Examples of QA/QC include model calibration, verification, and validation; use of expert judgment; and peer review. In evaluating the documents of the LOSLR Study, the committee evaluated how quality assurance/quality control was conducted, documented, and addressed to strengthen quality of the scientific products.

Criterion 3: Treatment of Error and Uncertainty

Error, uncertainty, and risk are inherent in scientific inquiry, so it is critical that the analysis and discussion of uncertainty is clear and forthright. Statistics is a tool commonly used to analyze uncertainty and error, and statistical convention is established in a language familiar to engineers and scientists alike. Error bars, confidence intervals, and statistical significance are just a few examples of how statistics convey uncertainty and error. Risk analysis is commonly used by engineers, and like statistics, risk analysis has convention and accepted practices associated with it (Haimes, 1981). The committee evaluated the LOSLR studies and models based on the criteria above and how well risk and uncertainty were documented and communicated.

CHARGE B: The LOSLR program science...will be reviewed in terms of the degree to which...the models [and studies] sufficiently and appropriately integrate and display the key information needed for a comprehensive evaluation and understanding of the tradeoffs for selecting among the candidate Regulation plans.

The committee interpreted Charge B to mean evaluating (1) whether the empirical study results and internal models of the SVM were integrated such that the SVM could be used to reliably evaluate various candidate regulation plans and (2) whether the SVM was effective in displaying key information to an external audience such as the NRC committee, review scientists, and the public at large.

The LOSLR Study structure depends on a series of integrated models that use input from empirical studies (Figure 1-6). These models have a common endpoint in the SVM, so it is important for them to be integrated and compatible with each other in spatial and temporal scales. Because the SVM is both a technical and a communication tool, it is also important for the linkages among the models to be well-explained and technically documented. Charge B ensures that the NRC review addressed questions of integration, resolution, documentation, and communication. Scientific criteria 4, 5, 6, and 7 (see Box 1-2) were used to address Charge B.

Criterion 4: Linkages and Feedbacks Among Related Studies and Models

The LOSLR Study structure (Figure 1-6) shows some of the linkages among the studies and models used in the SVM. Clear explanation of these connections is important to convey understanding of how the SVM works. Similar to model validation and verification for stand-alone models, the linkages and feedbacks between models within the SVM should be established; it is also important to ensure adequate resolution among the different pieces, i.e., criterion 5.

Criterion 5: Spatial and Temporal Resolution and Scaling

Evaluating candidate regulation plans in a complex system like the LOSLR basin involves data across a wide range of spatial and temporal scales. The resolution of the models and empirical studies, for those cases where empirical study results are used as model inputs, must be managed with care. Issues of scaling need to be considered in the analysis and documentation.

Criterion 6: Thorough Documentation

Thorough documentation of a body of scientific work will provide an outside audience with an understanding of (1) what was undertaken and accomplished; (2) the methods used; (3) any rationale behind an unconventional or unexpected method or approach; and (4) limitations of the methods used or results generated. This range of information can facilitate understanding and increase confidence in the scientific results and their policy applications, even when the documentation speaks to limitations on the science itself. Well-documented scientific work can also help to identify ways that methods or analyses can be improved in the future or monitoring activities related to the work.

While the value of thorough documentation holds true for all scientific work, it is especially important for multi-disciplinary work such as the LOSLR Study. In the LOSLR case, documentation is needed to understand how individual studies are conducted; how scientific results are incorporated into integrated models; and how models are used within the SVM. Furthermore, the SVM is a planning and modeling tool that, while not new, is still untested in most of its applications. The LOSLR Study could showcase the full range of SVM applications in a complex, multi-disciplinary water resources management project. Thorough documentation will aid in communicating the scientific and policy information to the many stakeholders in the LOSLR and to any other communities confronted with similar water resource management challenges.

Criterion 7: Effective Scientific Communication

Communication depends on one party disseminating clear information and another party receiving it in understandable terms. Effective communication of scientific information may vary depending on the scientific sophistication of the audience, and it is central to the Study's objectives and purpose. The LOSLR Study, by design, includes scientists from a range of disciplines, policy makers with a range of scientific and technical competence, and the general public. All involved parties are interested in the information that will come from the LOSLR Study, and effective communication of the scientific aspects is critical to the success of the Study.

> **CHARGE C: The LOSLR program science...will be reviewed in terms of the degree to which the models and reports are sufficient and appropriate to evaluate the various regulation plan options (RPOs) and impacts of changes in water levels and flows.**

Charge C was interpreted by the committee as evaluating whether the LOSLR Study components collectively provide appropriate and sufficient foundations to (1) evaluate, select, or eliminate regulation plans and (2) estimate impacts of water level fluctuations in the future. Criteria 8, 9, and 10 were used to address Charge C.

Criterion 8: Breadth of Study Scope

In the LOSLR Study, the water regulation plans will affect many aspects of the basin, such as hydropower, navigation, economic, ecological, and political interests. Thus, the LOSLR Study elements must represent a similar range of considerations. Using this criterion the NRC review evaluated the scope of LOSLR Study materials relative to the expected range of impacts in the basin.

Criterion 9: Balance Between Scientific and Practical Professional Approaches

Relationships between the scientific and practical professional communities are inherent in the LOSLR Study. The Study commissioned scientific research to help formulate and evaluate water regulation plan options, and the considered regulation plans and criteria affect some of the scientific questions and methods used. The LOSLR Study aims to provide the IJC with the scientific and management information that it needs to decide upon a regulation plan. This combination of accepted scientific and practical professional approaches often needs careful thought to remain operational. While scientists may prefer more time and resources to gain greater confidence in their individual studies, the water management community needs to prioritize and connect work conducted in different fields and approaches, and with limited time and resources.

There are many practical challenges involved in synthesizing vast amounts of information in a collaborative, negotiated planning process. Scientific studies may provide excessive detail that overwhelms the decision making process. Overly technical material may exceed human capacity to comprehend and evaluate a multi-criteria problem, with the unintended consequence of obscuring policy options or inhibiting decision making. Conversely, practical professional approaches may oversimplify complex phenomena and uncertainties. In these cases, the plan for linking scientific and practical professional aspects of studies and models needs to be clear.

A well-defined relationship between scientific and practical professional approaches will help scientists to streamline their efforts and policy makers to commission and use scientific studies and models. The water management community needs scientific information about physical and economic consequences in forms that are manageable, accessible, and useful to participants in the LOSLR type of collaborative, open, negotiation process. This relationship between the scientific and practical professional aspects of the LOSLR Study is important in determining the overall appropriateness and sufficiency of the LOSLR work to make decisions about regulation plans.

Criterion 10: Identification of Future Study Needs

The LOSLR basin has a long history of adapting regulation plans to changing conditions. The transition from Plan 1958D to Plan 1958DD is a good example of how a plan may need to adapt over time because of dynamic variables in the basin. Thorough documentation (Criterion 6) should also include a description of the limitations of current knowledge and identification of future study needs. Given that knowledge of the basin's ecology, political structure, and hydrology will change over multiple decades, it is important to outline how the regulation plan that comes from the 2005 LOSLR Study can adapt over time and build upon the science and efforts of the LOSLR Study. Future study needs may include new and better science, increased model sophistication, and various types of adaptive management.

THE NRC/RSC REPORT

Overall Aim of the Report

This report is produced in the spirit of supporting the IJC in its deliberations on the results of the LOSLR Study. The report recognizes the difficulty of using science for policy decision making purposes in a large-scale, water resources planning project with complex ecological, economic, and social considerations. The LOSLR Study has sought a comprehensive approach to water resources planning that is strongly encouraged. This type of large-scale, multi-faceted work is becoming more common as the connections among the economic, ecologic, and political aspects of hydrologic systems become more intricately entwined and better understood. Still, few precedents exist that can serve as templates for integrating multiple cross-disciplinary empirical studies and models or for communicating complex information to the public and policy decision makers alike. The LOSLR Study has pursued these aims, and this report, while praising the scope, breadth, and concept of the LOSLR work, makes constructive criticisms and recommendations to strengthen the scientific, technical, and integrative aspects of the work to ensure the success of the LOSLR Study.

Structure of the Report

The Shared Vision Model (SVM), Flooding and Erosion Prediction System (FEPS), and environmental studies are evaluated in the following chapters. Each of these evaluative chapters begins with an introductory section that outlines what was undertaken in the study, its accomplishments and its limitations. The second section of each chapter evaluates the scientific or modeling work presented to the NRC for review. Each chapter evaluation is presented in terms of Charges A, B, and C of the statement of task (see Box 1-1) and is reviewed using the ten evaluation criteria as described above. Each chapter closes with a brief summary of the findings and recommendations.

Chapter 2 describes and evaluates the SVM. The SVM is discussed first because it provides structure and context for the other topics of the committee's charge. Chapter 3 describes and evaluates the FEPS model. Chapter 4 describes and evaluates wetlands, SAR, and the IERM as a suite of work that represents the environmental sciences in the LOSLR Study. Finally, Chapter 5 reviews the main findings and recommendations of the report. It also offers a perspective on how the work initiated in the LOSLR Study can be advanced over multiple decades using adaptive management to adjust the regulation plan in response to dynamic changes in the LOSLR Basin and new scientific understanding.

REFERENCES

Canadian Hydrographic Service. 2005. Water Level Bulletin. Boundary Waters Treaty, 1909. Monthly. Available on-line at *http://chswww.bur.dfo.ca/danp/Graphs/bulletin.pdf.* Accessed July 2005.

Clinton Edmonds and Associates. 2002. Lake Ontario and St. Lawrence River-Changes in the Institutional Structure and Their Impact on Water Levels, 1950-2001. Report to International Lake Ontario-St. Lawrence River Study Board. Ottawa, CA: University of Ottawa, Institute of the Environment.

Environment Canada. 2005. Level News. Monthly. Available on-line at *http://www.on.ec.gc.ca/water/level-news/intro-e.html.* Accessed July 2005.

Haimes, Y. (Ed.). 1981. Risk/Benefit Analysis in Water Resources Planning and Management. New York, NY: Plenum Press.

IGLD (International Great Lakes Datum). 1985. The International Great Lakes Datum of 1985. Available on-line at *http://chswww.bur.dfo.ca/danp/igld1985.html.* Accessed July 2005.

ISLRBC (International St. Lawrence River Board of Control). 2004. Available on-line at *http://www.islrbc.org.* Accessed July 2005.

NOAA (National Oceanic and Atmospheric Administration). Great Lakes Environmental Research Laboratory. 2005. Great Lakes Water Levels. Available on-line at *http://www.glerl.noaa.gov/data/now/wlevels/levels.html.* Accessed August 2005.

USACE (U.S. Army Corps of Engineers) Detroit District. 2005. Great Lakes Water Levels. Available on-line at *http://www.lre.usace.army.mil/greatlakes/hh/greatlakeswaterlevels.* Accessed July 2005.

Werick, W. 2005. Presentation at to the NRC Committee to Review the Lake Ontario-St. Lawrence River Studies. June 14, 2005. Niagara-on-the-Lake, Ontario, Canada.

Chapter 2
The Shared Vision Model

This chapter discusses the Shared Vision Model that includes a planning process, a simulation model, and a suite of associated models. The first section of the chapter introduces the Shared Vision approach including its planning framework and modeling efforts. The second section of the chapter evaluates the Shared Vision Model, beginning with its scientific qualities, followed by its integration and display of key information, and its overall sufficiency for informing decision-making. Documents provided for this review included:

- ***SVM 1*** – Evaluation Process Overview – a two-page broad description of the Shared Vision Model;
- ***SVM 2*** – Shared Vision Model for the Design and Evaluation of Alternative Regulation of Lake Ontario Releases into the St. Lawrence River – this document includes the "Toronto Board Room April 24.xls" spreadsheet file and 83 linked files. The linked files include:
  - Contextual Narratives that describe scenarios of economic impacts on commercial navigation, environmental effects, recreational boating and tourism, land use along the Lower Saint Lawrence River, and hydropower;
  - Regulation Plan descriptions; and
  - Approximately 70 Performance Indicator fact sheets;
- ***SVM 3*** – SVM Documentation Report – a six-page description of the structure of the model.

The contextual narratives were provided as background rather than for review, along with a document on "Preliminary Criteria and Metrics for the Plan Formulation and Evaluation Group" and an "Informal Decision Tool Program," a spreadsheet file that provides an example of multi-attribute decision analysis. Staff members of the LOSLR Study made slide presentations and engaged in a three-hour question and answer period on the Shared Vision Model.

Shared Vision Planning and Modeling in the LOSLR Study

The LOSLR Study Board describes the Shared Vision Model (SVM) as "a unified computer simulation of the system" (SVM 1, p. 1; Werick, 2005). Its purpose is to evaluate previously formulated plans and alternative regulation plans as they are formulated. It is designed to be used by technical experts and stakeholders in a collaborative planning process for which its advocates have coined the term

"Shared Vision Planning." An article reproduced on the U.S. Army Corps of Engineers' Institute for Water Resources website describes shared vision planning as:

> ...deliberative, inclusive decision-making processes as the forum in which to debate how water resources will be used among competing ends. What is unique about shared vision planning, however, is how analytical technical expertise and analysis is integrated into a collaborative planning process. Through a structured planning process, an analytical computer model of the water resource system, called a shared vision model, is constructed with the participation of stakeholders. The shared vision model is designed to be used by stakeholders themselves to develop a mutually satisfactory water supply plan... (Stephenson, 2003, p. 1)

The U.S. Army Corps of Engineers introduced this approach to shared vision planning in its 1995 *National Study of Water Management During Drought* (IWR, 1995). The approach consists of three elements: (1) a structured planning process, similar to that described in the *Economic and Environmental Principles and Guidelines for Planning Water and Related Land Resources* (U.S. Water Resource Council, 1983) that states objectives and identifies problems and opportunities; formulates and evaluates alternatives; and displays effects of each alternative on each of the stated objectives; (2) a collaborative, negotiation process involving stakeholders; and (3) use of scientific and technical information through the use of computer models that accounts for the complex linkages among various elements of the water resource system under investigation (Stephenson, 2003). This last element of Shared Vision Planning is the Shared Vision Model.

The LOSLR *Plan of Study* is the official document that authorizes and defines the scope of the study (BACK 2). It was prepared by the IJC and approved by the two governments. The *Plan of Study* contains directives that are consistent with the three elements of a structured planning process, stakeholder collaboration, and computer simulation models described by Stephenson (2003). The *Plan of Study* mandated the use of a multiple-criteria rational planning model, similar to the one in the *Principles and Guidelines,* which guides water resource planning by federal agencies in the United States (U.S. Water Resources Council, 1983). The *Plan of Study* directed the LOSLR Study to investigate existing criteria within the IJC's Orders of Approval, and to determine the requirements for establishing new criteria for improved operation of the Lake Ontario-St. Lawrence River system. Existing criteria were primarily related to economic effects and functional requirements, particularly of hydropower, commercial navigation, and flooding. In referencing the 1996 Scope of Work, the *Plan of Study* emphasizes the need to consider wetlands, other environmental factors, and recreational boating interests, which had not been addressed in the original regulation plan. The LOSLR Plan of Study called for formulation of alternative regulation plans and evaluation of them with respect to multiple criteria, including both existing and new criteria that may be added. These directives comprise a structured planning process, which is the first element of the shared vision approach.

The *Plan of Study* mandated the formulation and evaluation of regulation plans using a collaborative, negotiation process involving a wide range of stakeholders. This process is the second element of the shared vision approach, it provided for creation of the LOSLR Study Board and assigned the Study Board overall responsibility for the Study. The Study Board's 14 members, half from the U.S. and half from Canada, are predominantly government personnel, academics, other technical experts, and Native peoples. The Study Board was further authorized to establish bi-national work groups to support its work. The *Plan of Study* (BACK 2, p. 13) emphasized that:

> Public consultation is critical to the assessment of plan criteria. It is recognized that progress in addressing water levels issues is dependent in large part on public understanding that most proposed solutions could have consequences for others. To achieve this understanding, it is recommended that the major interests and the relevant public be involved directly in the studies, by the formation of an Interest Advisory Group… The continuous involvement of all interests throughout the criteria review process is critical to the success of the study.

The *Plan of Study* (BACK 2, p. 76) also directed that computer simulation models would be used, i.e., the third element of Shared Vision Planning. It stated:

> The evaluation of Lake Ontario regulation plans, the practicality of proposed criteria, and the hydrologic impacts on the interests, require computer simulation of water levels and flows of the Great Lakes–St. Lawrence River system downstream as far as Trois-Rivères, Quebec. …While separate computer models exist for different parts of the system, and substantial progress has been made in the development of a coordinated routing and regulation model for the upper lakes, additional work is required to develop and integrate Lake Ontario regulation plan(s) and St. Lawrence River components into the model to simulate water levels and flows of the entire system (including the Ottawa River and other tributaries).

The *Plan of Study* even specified the time step for the simulation: because Lake Ontario outflows are regulated on a weekly basis, the computer simulation would use a quarter-month time-step. In this specification is the recognition that effects dependent upon shorter time periods would not be captured by models using the quarter-month time-step. Therefore, the *Plan of Study* called for development of a supplemental, two-dimensional hydrodynamic model for the St. Lawrence River upstream of Cornwall-Massena from near Kingston to the Moses-Saunders dam to simulate short-term effects on commercial navigation, recreational boating, environmental questions, hydropower, and ice formation.

The Shared Vision Planning approach has been used in other venues prior to its use in the LOSLR basin, including the following large-scale studies : (1) *Reservoir Operating Plan Evaluation (ROPE) Study for the Mississippi Headwaters* (Cardwell et al., 2004); (2) the *Rappahannock River Basin Commission Water Supply Planning Project*

(Connor et al., 2004; Werick, 2000); (3) the *Interstate Commission on the Potomac River Basin (ICPRB) Water Supply* (Hagen and Kiang, 2003); and (4) the *Alabama-Coosa-Tallapoosa-Apalachicola-Chattahoochee-Flint (ACT-ACF) River Basins* (Palmer, 1998). These examples represent a subset of a much larger literature on adaptive water resources management that has relevance for this study (e.g., NRC, 2004).

The LOSLR Study is commended for building upon this shared vision planning approach, and for addressing the three criteria described by Stephenson (2003). The LOSLR Study attempts to integrate these criteria in an innovative water resources planning process in the Lake Ontario-St. Lawrence River basin. Although integrated water resources planning is increasing in scope and application (e.g., NRC, 2004), the norm is still for scientific and technical analyses to be conducted separately from one another and from various facets of public decision-making processes and for models to lack transparency. These issues will be discussed further in relation to the Shared Vision Model, which is introduced below.

Shared Vision Modeling

As noted earlier, the LOSLR Study uses the term Shared Vision Model to describe a (1) planning process, (2) suite of sub-models, and (3) core simulation model. This section describes the computer modeling. The structure of the SVM consists of the suite of models shown in Figure 2-1 and described in general terms in the LOSLR Documentation Report (SVM 3).

The heart of the SVM is a computer model (the "STELLA" model in Figure 2.1) that simulates behavior of the Lake Ontario-St. Lawrence River system to Batiscan, downstream of Montreal, for any given regulation plan and time series of inflows. The STELLA model gets its name from the fact that it was written in the STELLA computer language. A simplified flow diagram of the simulation model as the committee understood it is shown in Figure 2-2.

The decision or control variables in the model are given in a Plan of Regulation that governs releases to be made from Lake Ontario through the Moses-Saunders Dam under various system inflow conditions. That time series of releases from the dam is then used to route a time series of quarter-month system inflows through a water balance model for the Lake; the outputs are a time series of lake and river levels. Downstream tributary flows are added to the dam releases at points of confluence to calculate time series of river flows and stages at various locations.

Flows are used within the STELLA model to calculate lake elevations, power, and navigation benefits, which are forwarded to three other programs that evaluate impacts on coastal, environmental, and recreation systems. Results from all four models are then presented in a "Board Room" spreadsheet.

The Shared Vision Model also includes several research components that supported development of the simulation model and related calculations of economic and environmental effects. Hydrologic and hydraulic (H&H) research activities

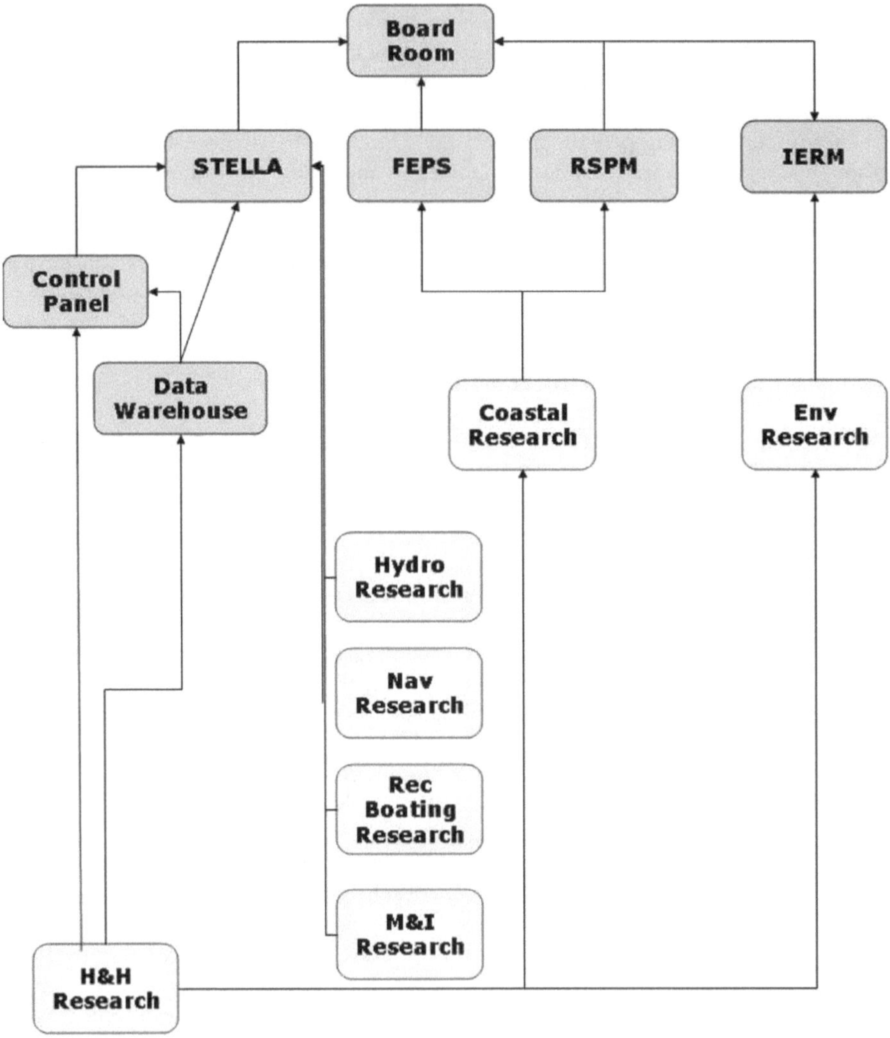

FIGURE 2-1 Structure of the shared vision model. SOURCE: Werick (2005).

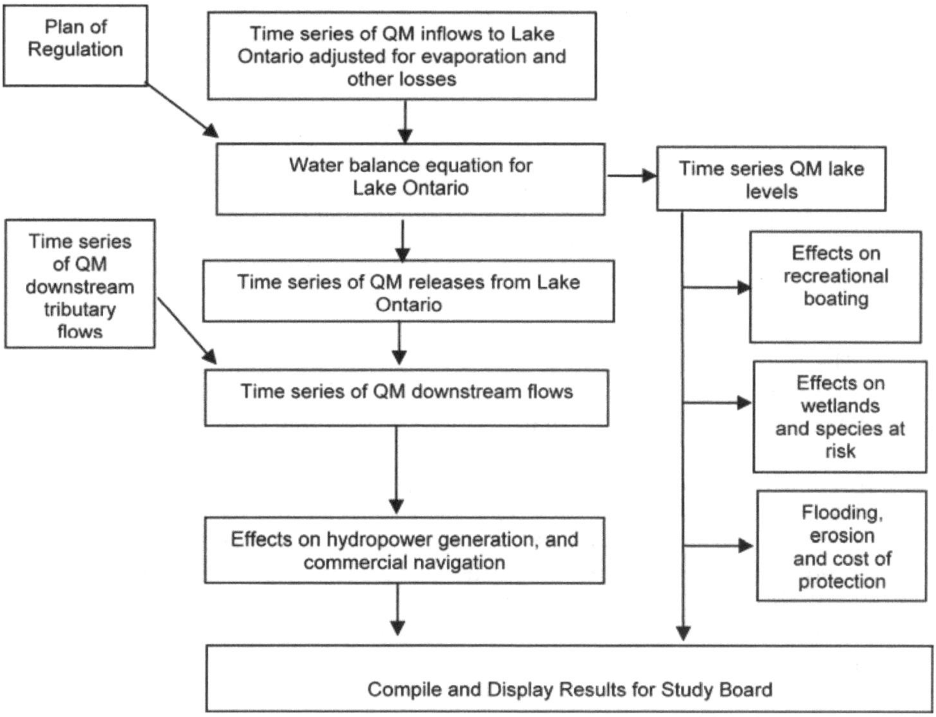

FIGURE 2-2 Committee interpretation of the Shared Vision Model.

produced time series of quarter-month system inflows adjusted to account for lake evaporation and other losses. Time series were generated for historical conditions, for historical conditions adjusted for climate change, and for stochastic series intended to represent probabilistically a wide range of possible future conditions. The hydrology and hydraulic research also produced estimates of ice and channel roughness and changes in temperature that may result from alternative plans. The Committee received a report on Lake Erie outflows (Fan and Fay, 2003) but did not receive other documentation of the time series of inflows to Lake Ontario used in the LOSLR study, and was not charged to review the hydrologic and hydraulic work.

Research was also undertaken to establish relationships between water levels and flows and important effects of regulation, namely municipal and industrial water supply (M&I), recreational boating (Rec Boating), the environment (Env), commercial navigation (Nav), coastal flooding and erosion (Coastal), and hydropower production (Hydro). The relationships established through these research activities were used to build three sub-models. One is the Flooding and Erosion Prediction System (FEPS), used to predict flooding and erosion along the coasts of Lake Ontario and the river above the regulating dam. FEPS takes a time series of lake levels from the STELLA model (passed through the Board Room) to calculate a time series of (1) erosion of developed land that has no shoreline protection, (2) costs of maintenance and repair to land that has coastal shore protection structures, and (3) flood damages. A second sub-model, the River Shore Protection Model (RSPM), calculates impacts and costs of erosion of existing shore protection below the Moses-Saunders dam. The third sub-model, the Integrated Ecological Response Model (IERM), uses relationships established in the environmental research program to estimate a large number of environmental performance indicators resulting from a given regulation plan and a given time series of lake levels. FEPS and IERM are reviewed in chapters 3 and 4, respectively; the RSPM was not subject to this review.

The Board Room is a spreadsheet file. It takes the simulated time series of hydrologic attributes from the STELLA model and sends them to the FEPS, RSPM, and IERM models to estimate effects on performance indicators. It then receives results from those sub-models, and prepares statistical summaries as "…requested by decision makers and stakeholders, including average annual, worst case, persistence of bad performance, etc." (SVM 3, p. 2). The Data Warehouse stores alternative time series of inflows for use in the SVM. The Control Panel is described as "…a spreadsheet file that helps the SVM user design new regulation plans…" (ibid, 4). It also provides the user with options from which to select a time series of inflows stored in the Data Warehouse as an input to the SVM. The Data Warehouse and Control Panel were not included in this review.

Brief Description of Regulation Plan Options

The primary purpose of the LOSLR study is to formulate regulation plan options for the releases from Lake Ontario and the upper St. Lawrence River through the Moses-Saunders dam. The Board Room includes information on seven plans formulated and evaluated in detail. Brief descriptions of the plans included in the Board Room are:

- Plan 1958 D – the plan that went into operation in October of 1963;
- Plan 1958 D with Deviations (Plan 1958DD) – Plan 1958D with deviations that the Control Board has been using in recent years to cope with outcomes it considered to be unacceptable;
- Plan A (also known as Cornell V) – described by the Study Board as "the balanced economic plan," (Available on-line at *http://www.losl.org/PDF/17062005_NewsR-EN.pdf*. Accessed August 7, 2005).
- Plan B (also known as NaturalLy) – described by the Study Board as, "the balanced environmental plan" (Available on-line at *http://www.losl.org/PDF/17062005_NewsR-EN.pdf*. Accessed August 7, 2005).
- Plan D – described by the LOSLR Study Board as, "the blended benefits plan" (Available on-line at *http://www.losl.org/PDF/17062005_NewsR-EN.pdf*. Accessed August 7, 2005).

- Plan E – The Pre-project Plan with Ice Limits incorporates historic pre-project winter ice retardation effects on pre-project flow.
- Plan 1998 – versions of the plan proposed in 1998 that was rejected by the IJC.

Outcomes from the current "Plan 1958 D with Deviations" (or Plan 1958DD) were used as a baseline against which other plans were compared. Although this review does not comment on the specific regulation plan options listed above, it does note here and discuss below that the plan formulation process appears to have focused on specifying release criteria to accommodate *future* hydroclimatic scenarios under *current* economic and environmental conditions. Plan formulation does not appear to have focused on provisions for monitoring, testing, evaluating, and, if necessary, adapting plans and criteria to changing conditions and priorities.

EVALUATION OF THE SVM OVERALL FINDINGS AND ACCOMPLISHMENTS

The overall strength of the SVM is its ability to support multi-stakeholder decision-making with data-driven comparisons of tradeoffs among the economic, environmental, and social effects of regulation plan options. The inter-disciplinary scope of inquiry of the SVM and its inclusive approach will likely serve as an important case study for other large-scale, multi-interest, water resources planning projects. These SVM strengths, however, are undercut by the poor model documentation that leaves unanswered many questions of the rationale behind unexpected modeling choices, methodology, treatment of model error and uncertainty, and whether the SVM in fact is integrating scientific and economic components in a technically and scientifically sound manner. A more detailed evaluation of the SVM follows that corresponds with the three charges in the statement of task: scientific foundations, integration and display of key information, and overall appropriateness and sufficiency of the SVM to inform regulation plan decision-making.

EVALUATION OF SCIENTIFIC METHODS, ASSUMPTIONS, AND SUPPORTED FINDINGS

Charge A of the statement of task focuses on the extent to which the studies and models reflect reasonable scientific methods, assumptions, and supported findings. Three evaluation criteria are used to assay the scientific foundations of the SVM: empirical foundations, quality assurance, and treatment of risk and uncertainty.

Criterion 1: Empirical Foundations

The primary purpose of the SVM is to convey how changes in lake level and river flow will impact a variety of other dependent conditions in the LOSLR system (e.g., wetlands, recreational boating, flood damages to property, etc.). To fulfill this purpose, the SVM is designed to draw from several empirical bases, the majority of which are either not subject to this NRC review or are reviewed in subsequent chapters of this report. Therefore, this section focuses on three specific empirical issues: lake level calculations, use of regression analysis for modeling the St. Lawrence system, and assumptions about related processes and trends within the STELLA model.

Single Lake Level Calculation

Water levels and flows in the LOSLR system are generated using hydrologic time series input from a limited number of points in the lake and the river. The input flows used by the SVM were generated from (1) historical records adjusted for current diversions, (2) historical flows adjusted for climate change, and (3) stochastic flow series intended to probabilistically represent a wide range of possible future conditions. Most input flow evaluations were based on 101 years of record (4,848 quarter-months). A stochastic sequence of flow for 50,000 years was generated and, from that sequence, 101-year records of very wet and very dry centuries were selected to evaluate plan performance under extreme conditions. IJC documents do not discuss how flow sequences were generated or why 101 years was chosen as the length of record. Stochastic time series would normally include multiple realizations that can be used to generate probabilistic model outputs, but documentation did not indicate that multiple realizations were used in the LOSLR application.

The STELLA model produces a single water elevation value for Lake Ontario for each quarter-month time period. This STELLA model value must be accurate for the SVM to function reliably because it is used in other models as input for wetlands, fauna, erosion, and bathymetric calculations. One problem in using a single value in the quarter-month time step is that Lake levels can be more dynamic than this value can capture. For example, the average change in water elevation of Lake Ontario may be as much as 20 cm in one week (*http://www.glerl.noaa.gov/data/now/wlevels/levels.html*). This large range in water elevation values is nontrivial because erosion and ecological models use the average values to calculate coastal and environmental responses, respectively. Although some of the sub-models generate variable water level conditions from the average values, any inaccuracies in the average water level values would propagate through the erosion and ecological model outputs and the SVM. It is therefore surprising that the SVM documents do not report on how the accuracy of average water level values is determined, or how sensitive the other models are to the accuracy of this value.

The problems of relying on a single water elevation value for Lake Ontario can be summarized in three points. First, water level data are available at multiple locations on Lake Ontario and at a finer time step than quarter-month period. The decision to calculate a single model output for a larger area and longer time period needs better justification. Second, the coastal and environmental models use these less precise single value quarter-monthly outputs to calculate finer-scale water level and wave processes in different areas of Lake Ontario and the Upper St. Lawrence River. Third, those finer-scale hydrologic scenarios are subsequently used to calculate a chain of environmental, coastal, and economic impacts. **Thus, the decision to use a single lake-wide water level value needs fuller explanation and an analysis of the accuracy of a single water level value in the SVM and submodels.**

Use of Regression for Modeling the St. Lawrence System

The STELLA model calculates river water levels and flows at a limited number of points based on regression equations (Fan and Fay, 2001, 2002, 2003). Regression analysis is straightforward and inexpensive, which may be the reasons it was chosen for use in the LOSLR work.

Several questions remain unanswered with respect to regression analysis in the SVM. First, why were regression analyses used to a greater extent than conventional hydraulic models? Given that hydraulic models are empirically feasible, and were recommended in the *Plan of Study* (BACK 2), it is unclear why the SVM relied more heavily on a different and less accurate water level calculation method. Second, how were the regression analyses developed, and how accurate were the results? Regression is a common statistical tool that is described by standard statistical descriptors, such as r^2 and confidence intervals. These descriptors are not included in the SVM documents, and no measures of accuracy or reliability are presented to support the regression analysis approach. Without proper documentation of the development and accuracy of the regression analyses, and the relationships between the STELLA model and the hydrodynamic modeling, it is not possible to render a judgment about the adequacy or accuracy of the regression approach. In fact, regression could lead to inaccurate results, which spurs concerns about the use and application of regression analyses in the Shared Vision Model. **The ways that lake and river hydraulics models and regression analysis were used with the SVM need additional justification to defend their selection and application.**

Assumptions about Related Processes and Trends within the STELLA Model

The SVM has the ability to support multi-stakeholder decision-making using tradeoffs among economic, environmental and social effects of water level regulation. However, the reviewed documents suggest that related socioeconomic trends and their possible feedback on system operation were not incorporated into the SVM. For example, population, land uses along the coastline, and demands for commercial navigation and recreational boating were specified externally. For the most part, values for these variables were fixed at current levels such that it is not possible to determine the effects of water level and flow regulation on these processes. Also omitted from the SVM analysis are the impacts of changes in externally specified water demands and land uses on the economic and environmental performance indicators modeled under alternative regulation plans. **These deficiencies in socioeconomic analysis, forecasting, and scenario construction limit the long-term utility of the SVM, and the influence of dynamic socioeconomic factors on modeled output variables must be treated more fully.**

Criterion 2: Quality Assurance and Quality Control

Model verification and sensitivity analysis are commonly used to compute and report model accuracy. Sensitivity analysis determines model parameter effects on model results and assesses model uncertainty, but sensitivity analyses of the SVM were not presented in the review documents. IJC documentation (SVM 3) states that validation and verification of the SVM involve a system of internal, self-assessments by the Technical Working Groups responsible for the development of each sub-model. Questions in these self-assessments included the following (SVM 3, p. 5):

- "Is the underlying research trustworthy?"
- "Is this the information you [stakeholders] were looking for?"
- "Is this information being interpreted correctly?"
- "Is a 101 year average a misleading metric?"

Although these questions are important, they indicate a different approach to validation, and it is unstated as to why these types of questions were used in lieu of traditional model validation, verification, and sensitivity analyses that would quantify inaccuracies due to underlying computational components, incorrect model choice, inappropriate approximations, or poor parameter estimation. Expanding the validation process to include these types of more traditional methods of model verification would yield greater levels of confidence in the SVM and its sub-models. **To give the highest confidence that the SVM and its sub-models are accurate and perform as designed to meet model expectations, traditional validation and verification methods should be employed and the results should be presented for formal external peer-review.**

Criterion 3: Treatment of Error and Uncertainty

Treatment of uncertainty in the SVM raised questions similar to the treatment of model validation (Criterion 2). Overall, uncertainty aspects of modeling have not received adequate attention; a consistent approach is needed to deal with various aspects of uncertainty but is not presented in the SVM materials. An exception to this overall finding is the sound approach found in the Cornell V modeling procedure. In this procedure uncertainty about future inflows was accounted for by dividing historic inflow sequences into five categories from very dry to very wet (SVM 2, 2005). Probabilities were assigned to each category based on relative frequencies in the historic record. That process was repeated for each initial lake level and each future flow sequence. Results are used to find the release that minimizes the expected deviation of performance indicators from established targets and limits and are attempts to account for random uncertainties that affect water supply by including synthetic input data, e.g., "stochastic" and "climate change" series, in the analysis.

The SVM works via input/output connections among environmental, engineering, economic, and scientific data and models. These connections provide an opportunity to track error and uncertainty throughout the entire SVM structure (Figure 2-1). However, the SVM model and sub-models have been run deterministically and have not analyzed errors and uncertainty propagated through the modeling system to the outputs used to compare regulatory plans. The SVM Board Room displays scatter plots for some variables, and it has links for contextual narratives that address uncertainty in qualitative terms, but it lacks a comprehensive framework to assess uncertainty among its interacting models. **The LOSLR Study should inform decision-makers about the ways that uncertainties were, and were not, addressed in the Shared Vision Model, and provide a detailed discussion of their relevance for decision-making.**

Summary of Scientific Methods, Assumptions, and Supported Findings

In summary, the scientific foundations of the SVM present questions that will need to be addressed before determining whether the SVM is scientifically sound. Some of these questions can be addressed through more thorough documentation. Others require more substantial work to strengthen the scientific basis for using the SVM results in decision-making. Specific recommendations to improve the scientific foundations of the SVM include the following:

- Hydrologic time flow series need better documentation as to how they were composed
- The decision to use the quarter-month time step needs more thorough explanation, examination of its use and interpolation in environmental and economic models, and better defense of its accuracy
- The influence of socio-economic factors on output variables of interest to the stakeholders needs to be considered more fully
- Standard practices for model calibration, validation, and sensitivity analy-

sis must be implemented in the SVM system with better documentation and clearer explanation

- A process is needed to treat uncertainty in ways consistent with standards of modeling practice.

INTEGRATION AND DISPLAY OF KEY INFORMATION

The second charge in the statement of task probes how well the SVM studies and models integrate and present key information for use in the regulation plan decision-making process. Display and integration of information in the SVM were evaluated using these criteria: (1) linkages and feedbacks among related studies, sub-models, and external factors, (2) resolution of spatial and temporal scales, (3) thorough documentation, and (4) effective scientific communication.

Criterion 4: Linkages and Feedbacks among Related Studies and Models

Integration and linkages among the SVM sub-components may be the most important aspect of the SVM structure. The SVM is composed of several levels of information: empirical research data support a range of sub-models; these sub-models provide analyses and values for hydrologic, environmental, and economic variables; and the STELLA model and Board Room integrate output from other sub-models. This structure depends upon smooth integration among all of its components, strong linkages between the models and the Board Room, and feedback from exogenous factors, such as economic and demographic trends (Figure 2-1).

Internal and external problems are noted with SVM integration and linkages. Internal problems relate to linkages between the SVM and its sub-models and a lack of feedback between the SVM and other sub-models. For example, there are missing links between bathymetric models and environmental models (see Chapter 4 for environmental sciences modeling discussion), and the STELLA model quarter-month output is used in environmental models that were designed with daily time steps. The introduction of error due to the lack of model feedback is unexamined, but could be significant. The SVM and its sub-models need better integration to limit error and provide important feedback.

External problems relate most strongly to alternative economic and demographic scenarios that will be relevant for the water regulation plan that replaces Plan 1958DD. For example, the changing socio-economic context of the LOSLR system will influence demands for various services from the LOSLR system (e.g., municipal water supplies, recreation, navigation, ecosystem services, etc.), but the SVM does not incorporate these scenarios in its structure. Likewise, socio-economic factors relevant to water level regulation (urban runoff will affect wetland restoration and net benefits even if water levels are favorable) are not well described or incorporated in the model structure. These issues are briefly outlined in conceptual narratives, but they are not modeled. **More robust linkages among the SVM and its sub-models, and integration between empirical research, exogenous factors, and SVM models are needed.**

Criterion 5: Spatial and Temporal Resolution and Scaling

Variable water levels and flows within the LOSLR system need to be determined at appropriate temporal and spatial resolutions for the entire system in order to assess the effects of changes in those levels and flows. As discussed above, the STELLA model generates water levels and flows in quarter-monthly intervals. The Flooding Erosion and Prediction System, the Integrated Ecological Response Model, and other environmental sub-models use water-related information on shorter time-steps. These time-step differences were partially recognized at the formulation phase of the study, but the decision was made to proceed with a quarter-monthly model because Lake Ontario outflows are regulated on that basis. Some efforts were taken to mitigate effects of inconsistency among different time-steps. Information about a 2-D hydrodynamic model developed for the upper St. Lawrence River and details regarding its calibration and validation were not provided for NRC review (BACK2; LOSLR Study, 2005, p. 20). In some cases, algorithms were developed to superimpose shorter-term variability on quarter-month averages and to introduce spatial variability (Fan and Fay, 2001, 2002, 2003). Even so, the quarter-monthly time step creates problems of temporal resolution with other components of the LOSLR and SVM that operate on finer temporal and geographic scales than those generated by the STELLA model. The amount of error introduced by these differences in temporal and spatial scales between the STELLA model and other SVM sub-models is unknown. Too many critical processes, including environmental and shipping impacts, occur on shorter time frames, and the reconciliation does not seem adequate between the shorter time frames in some model subsystems and the quarter-month intervals in the SVM.

Criterion 6: Thorough Documentation

Documentation of the SVM created systemic problems for this NRC review. Few materials describing the SVM were presented for review. Of those materials, some reports were incomplete, and some of the modeling components were not operable. These documentation problems limited the review committee's ability to test and fully understand the design and operation of the SVM and its sub-models. One example of the documentation deficiencies can be seen in the main report on the SVM, a 6-page document entitled, "SVM 3 SVM Documentation Report." The last sentence of that report states, "…the following pages list the equations, equation documentation and embedded array data from the STELLA model, the main component of the SVM," but the report ends without listing the equations or documentation of the STELLA model.

The limited access to and poor documentation of the inner workings of the STELLA model raised questions about how the model is programmed and how it incorporates stakeholder participation. In addition, the accuracy of the resulting calculations could not be verified. **The equations that comprise the SVM models need better documentation to convey the model structure and logic.**

The SVM Board Room includes descriptions for four regulation plan options and analyses for several others. During the time of the NRC review, the IJC was presenting three possible regulation plans for public comment, referred to as Plans A, B, and D. Although this review does not comment on the plans themselves, it does note that plan descriptions vary in level of documentation on the Board Room spreadsheet. For example, Plan A has a full description of the optimization model used to find the current quarter-month release by minimizing the percentage deviations from specified upper and lower targets for the performance measures over the next year (48 quarter-months). It indicates that the target levels come from the Public Interest Advisory Group presentations at public meetings in summer 2004 (see Cornellreport.doc and PIAG_Yr4_Draft_17.ppt). By comparison, the two-page description of Plan B provides limited insight into how that plan was derived because details were not included, and Plan D is not clearly described at all (Toronto Board Room April 24.xls [SVM 2]). **Any plans presented for consideration should be described in a detailed, consistent format.**

The state of SVM documentation raises serious questions as to how effective the SVM can be in generating a shared vision. One of the difficulties with plan documentation, and SVM output documentation more generally, is that the models were continuously adjusted over the course of the NRC review. These refinements made it difficult to keep up with current model versions, specifications, output, and documentation. The incomplete documentation is particularly troubling because some aspects of the SVM are very complex and need explanation to be transparent and comprehensible to outside, interested parties. **More thorough documentation of the SVM is needed to convey transparency, specifically on the rationale for model choices, methods used, treatment of error and uncertainty, and resolution of spatial and temporal scales.**

Criterion 7: Effective Scientific Communication

The primary venue for communicating information in the SVM is the "Board Room," a spreadsheet that displays the effects of regulation plans for visualization, deliberation, and further modeling decisions by the LOSLR Study Board. Through the Board Room, output from the Integrated Ecological Response Model, the Flooding and Erosion Prediction System model, and other models are compiled, displayed, and communicated. The Board Room displays information about the results of scientific studies and models, and to some extent how those results informed decisions about regulation plan formulation and evaluation in the SVM process. It thus provides a partial basis for assessing the appropriateness and sufficiency of the scientific studies for informing further decision-making. The Board

Room also houses and displays some background memos that have guided the LOSLR Study. Compared to other SVM materials reviewed, the Board Room offers a broader perspective on the appropriateness and sufficiency of the Shared Vision Model because it presents a wide range of comparative information on effects of regulation plan options on performance indicators.

The Board Room has a clearly organized home page consisting of five broad columns: Descriptions; Plan results; Levels and flows; Guidelines; and A Closer Look (at mitigation, climate change, and other issues) (Figure 2-3).

It also indicates the array of documents, tables, graphs, and diagrams developed to compare the potential effects of different regulation plan options, including the following information in Figure 2-3:

- Plan Descriptions (column 1)
- Criteria Metrics Description (column 1)
- Plan Results by Regions (Economic benefits and environmental benefits by region. These results are based in part on group workshops held in Montreal, Brockville, Burlington, and Syracuse) (column 2)
- Plan Results by Interest (coastal [flooding, erosion/shore protection], recreational boating, commercial navigation, environment, and hydropower stability and predictability) (column 2)
- Plan Results, including: environmental performance indicators results; economic performance indicators results; and social performance indicators results (the hydropower link was not live in the current version) (column 2)
- Study Board Screening Factors (a list of 21 factors points used by board members to evaluate each plan) (column 1)

By presenting water level effects in these manifold ways, the Board Room offers detailed and, in some measure, transparent insights into the types of information used to make comparisons among potential plans. Similarly, the Board Room presents many different ways of visualizing these data including:

- tables of impacts
- circular ratio plots of multiple performance indicators (Figure 2-4)
- 2D graphs of overall results [-1,-1 to 1,1] (Figure 2-5)

bar graphs.

Many of these displays are creative and useful. Some, like the circular ratio plots, allow comparison of a wide suite of performance indicators (Figure 2-5), while others present data aggregation problems that will be elaborated in chapter 4.

Although the Board Room is an excellent idea for communicating the results of the SVM modeling process to decision makers and stakeholders, it does not yet live up to its full promise. For example, the Board Room lacks an introduction on its homepage or a detailed user's manual, and it has some broken links. More substantively, some of the broken links are on important and relevant aspects of the

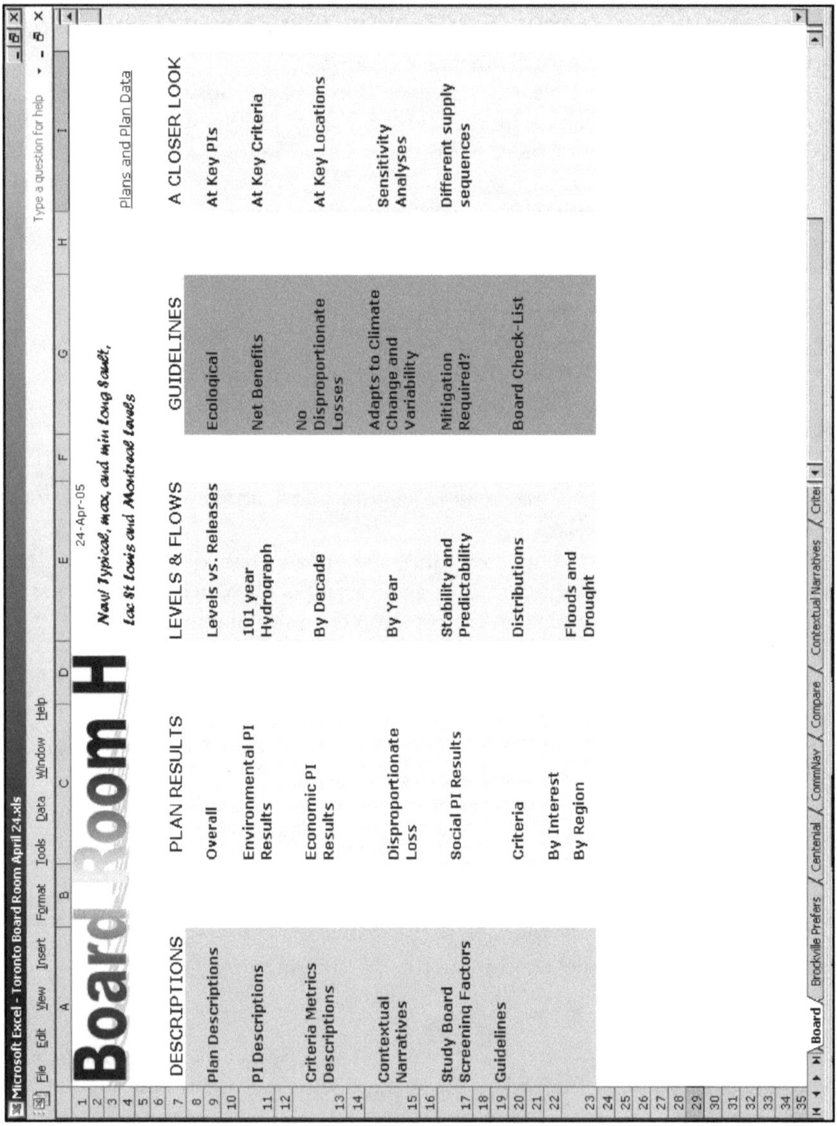

FIGURE 2-3 The Board Room interface. SOURCE: Board Room, Toronto Board Room April 24.xls (SVM 2).

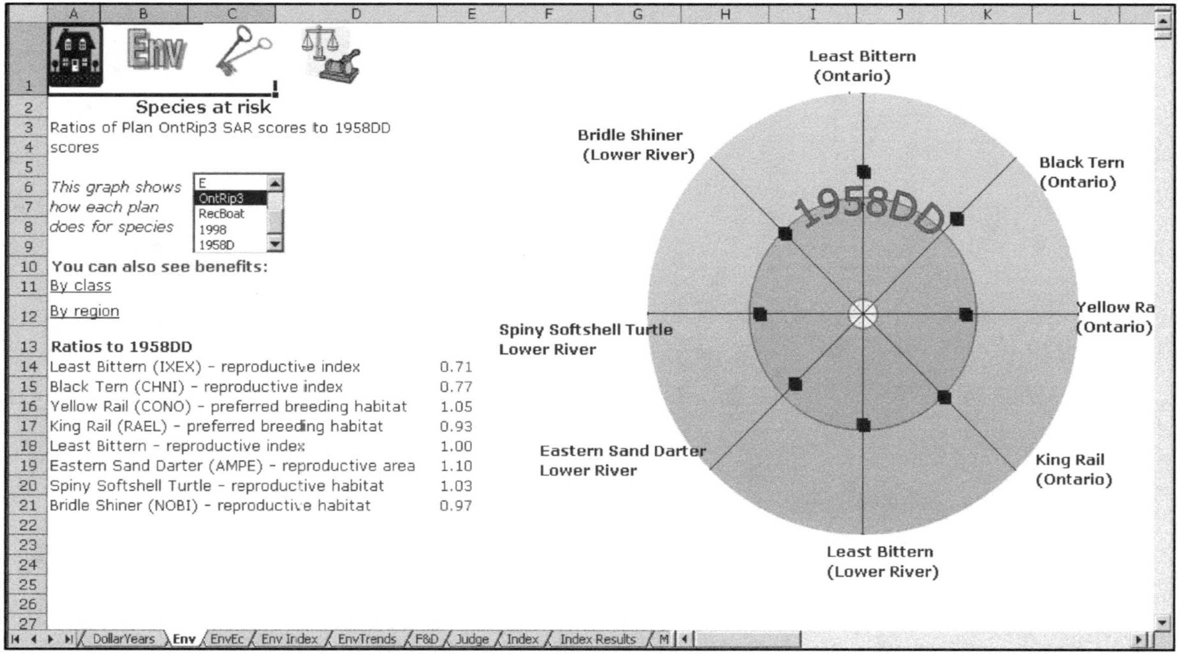

FIGURE 2-4 Circular ratio plots of multiple performance indicators in the Board Room. SOURCE: Board Room, Toronto Board Rocm April 24.xls (SVM 2).

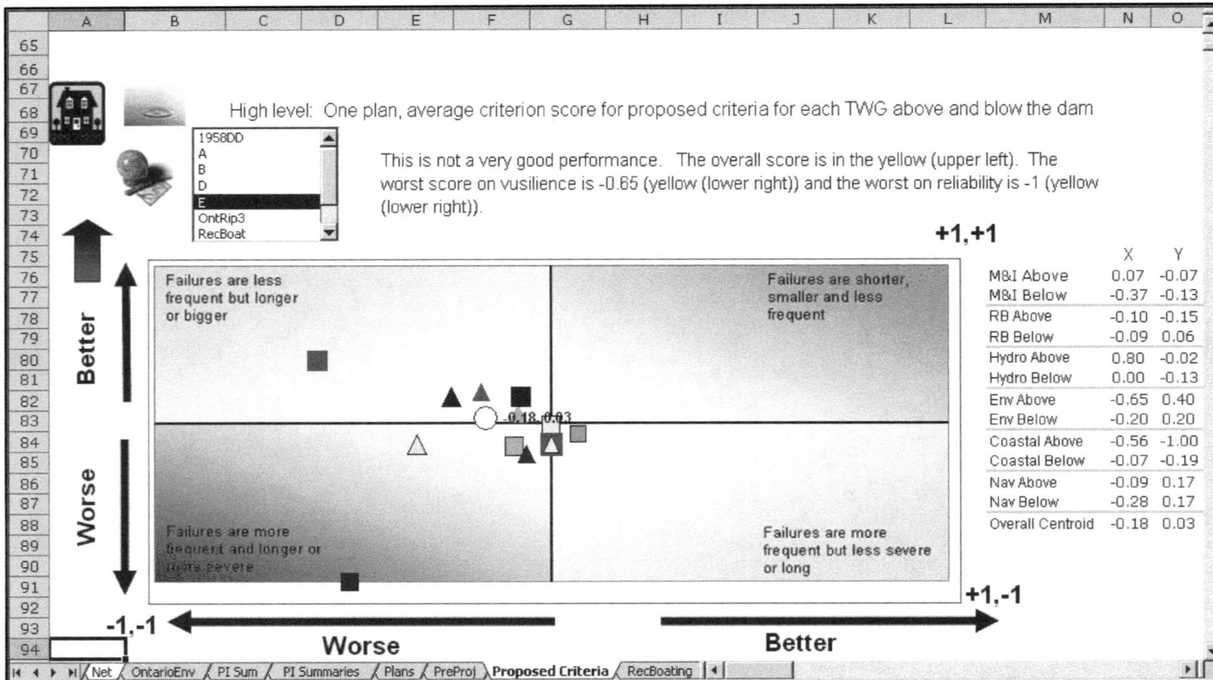

FIGURE 2-5 Frequency, severity, duration graph for multiple criteria presented one plan at a time.
SOURCE: Board Room, Toronto Board Room April 24.xls (SVM 2).

SVM including climate change, mitigation, sensitivity analyses, and water supply sequences. In addition, it lacks an archiving function of the regulation plan options that were formulated and evaluated. This archive function would improve the transparency of the SVM process, and enhance its long-term value for water management. **Therefore, the Board Room should present information on the uncertainty of the SVM data, analyses, and comparisons; and it should "archive" model runs that revealed model weaknesses and regulation plan option strengths and weaknesses.**

Summary: Integration and Display of Key Information

Four evaluation criteria were used to assess how well the SVM integrated and presented key information: model linkages, spatial and temporal scale resolution, model documentation, and effective scientific communication. Overall integration among internal SVM components and feedback with external factors, such as socio-economic factors, were found to be lacking. Model linkages were found to be problematic both for internal linkages within SVM components, and for external feedback of exogenous factors (such as socioeconomic factors) into the SVM. Spatial and temporal scaling present challenges for the SVM because the STELLA model produces water levels and flows per quarter month time step while many of the SVM sub-models operate on a much shorter time step. In addition, disparities in reporting economic and environmental indicators constrain stakeholders and decision-makers' ability to weigh these outputs intuitively. Inadequate documentation of the SVM was systemic. It was manifested in: (1) lack of information about the equations incorporated in the STELLA model; (2) incomplete SVM models and sub-models that precluded off-site execution; and (3) inadequate description of the development, selection, and elimination of alternative water regulation plans. The inability to determine what was done, how the SVM was constructed, and the rationale behind many unconventional approaches reflects a failure of the SVM to achieve the overall goal of transparency of the modeling process.

Recommendations to improve the integration and display of key information in the SVM follow.

- Linkages among SVM sub-models should be strengthened
- Feedback between exogenous factors needs to be strengthened in the SVM structure
- In cases where temporal and spatial scales are different, efforts should be made to improve compatibility, or to more clearly present the error and uncertainty created by scaling issues
- SVM documentation should be improved, at a minimum, in the following ways:
    - Better description of the SVM and its role in the development, selection, and elimination of alternative water regulation plans, and
    - Clearer rationale behind choices different from those of standard

practice for methodologies of model development and its use in plan formulation and evaluation.

OVERALL APPROPRIATENESS AND SUFFIENCY TO INFORM REGULATION OPTIONS

Charge C in the statement of task questions the extent to which the models and studies are sufficient and appropriate to evaluate the various regulation plans and the impacts of changes in water levels and flows in the LOSLR system. Three evaluation criteria are used to address this charge with respect to the SVM: breadth of inquiry and scope of the SVM, its balance between scientific and practical professional approaches; and identification of future studies needed to fill information gaps.

Criterion 8: Breadth of Study Scope

The SVM and its related planning process involved over 150 scientists, nine Technical Work Groups, a Public Interest Advisory Group, numerous stakeholder meetings, and a management structure of engineers and planners. It sought to expand the scope of water regulation planning to address hydrologic, environmental, and economic processes on timescales ranging from hours to decades and spatial scales ranging from meters to hundreds of kilometers. By historical standards in Great Lakes water regulation, it achieves an unprecedented breadth of scope in its coordination of various management aspects of a complex international water system.

The SVM aims to facilitate a transparent process to replace water regulation Plan 1958DD that includes inter-disciplinary approaches (as described in Breadth of Scope in Chapter 1). The inter-disciplinary approach was used to ensure that the selected plan best reflects the range of input received from the scientific, policy, and public communities. Range of choice issues, as related to the scope of the SVM approach, are identified as: 1) expanding the range of choice among regulation plan options; 2) analyzing the range of choice among regulation plan options; and 3) comparable treatment of environmental and economic effects of regulation plan options.

Scientific Basis for Expanding the Range of Choice

The SVM expanded the domain of water regulation to consider environmental, recreational, and coastal indicators, as well as established water uses. It also formulated and evaluated a large yet unknown number of regulation plan options that presumably expanded the range of choice. It did so through a multi-party stakeholder negotiation and evaluation process that used the SVM. Although that process does not fall within the scope of this review, several points are pertinent to the evaluation of the SVM.

In open session, LOSLR staff stated that participants were invited to submit plans for evaluation using the SVM, but they do not appear to have provided detailed technical guidance, directions, or tools to help stakeholders prepare candidate regulation plans. It is possible that in such multi-party negotiations, the parties' self interest would motivate them to propose plans that they favor; other parties would counter with proposals that they favor, and this could eventually lead to an exploration of the feasible space for solutions. In such a process stakeholders may propose alternatives, examine results, "learn" about how the system might respond, and iterate again. However, multi-party negotiation approaches may miss opportunities for joint gains and alternatives that expand the range of choice significantly beyond existing water regulation regimes.

The SVM Board Room includes mitigation measures, which constitute another important set of ways to expand the range of choice by reducing losses and increasing net benefits. Approaches for incorporating mitigation analytically in the SVM may involve non-structural as well as structural adjustments to water regulation impacts (e.g., flood damages, see FEPS 8 for an example of structural mitigation). They can involve (1) changes in a specific regulation plan that reduce a negative impact; (2) measures applicable to all regulation plan options; or (3) a combination of these two approaches. The Board Room includes a link labeled "Mitigation Required?" but that link is empty, so it is not clear if it encompasses the modeling opportunities noted above. **The LOSLR Study has examined a large number of alternatives that need to be documented to indicate the range of choice explored through the SVM.**

Scientific Bases for Evaluating the Range of Choice—The Possibility Frontier and No Disproportionate Losses

The process of evaluating and screening potentially attractive plans from the overall set of possible plans is critical to the success of the Study. The number of plans that were formulated through the 5-year Study was not presented for review, although anecdotal estimates indicate that over 100 plans were developed and evaluated. Several important steps in the SVM plan evaluation process remain unclear: (1) how it narrowed the universe of possible plans to the seven plans listed on the Board Room, (2) how it narrowed those plans to the three candidate plans under consideration, and (3) how the SVM was used for evaluating, screening and selecting these alternatives (see Leger et al., 2005 for a draft account).

The SVM would benefit from a clearer presentation about how it was used to evaluate the range of choice among regulation plan options. The Board Room contains much evidence of evaluation (e.g., tables and diagrams comparing impacts, and Study Board screening factors), but it is less clear how the Study Board evaluated the range of choice among the emergent alternatives (Wescoat, 1987; White, 1964). This process can be pursued in a number of ways, and the SVM took several approaches. The Board Room spreadsheet includes a list of 21 screening factors used by the Study Board. Although it does not provide supporting information about how those factors were applied to specific options, it does indicate the importance attached to different evaluation criteria. It is still not clear whether the SVM was used to identify the possibility frontier of feasible alternatives.

The Possibility Frontier. A "possibility frontier" of efficient solutions (Cohon, 1978; Goicochea et al., 1982) can be depicted as a curved boundary that that separates technically feasible plans (inside the boundary) from technically infeasible plans (outside the boundary). In an illustrative case of two objectives, improving the environment and increasing net economic benefits, plans on the boundary are efficient, in the sense that it is not possible to improve the performance for one objective without decreasing performance of the other (Figure 2-4). A possibility frontier can also be developed for a multi-objective optimization problem, where the boundary is a 3- or more-dimensional surface. In either case, the possibility frontier can be used as a guide for those who are charged with deciding among efficient plans. Although the SVM does not identify a possibility frontier of efficient alternatives, it does enable evaluation of plans that are relatively better or worse for different indicators, regions, and interests, which the Board Room displays in detail. SVM materials do not document the process employed to formulate plans that increase one set of benefits without decreasing others (i.e., Pareto-optimal type solutions), or increase benefits for some without causing disproportionate loss to others (though see Leger et al., 2005).

Disproportionate Loss Criteria. The SVM presents board members' screening factors in a level of detail that is commendable. The Study Board uses the criterion of no disproportionate losses, for instance, to eliminate water management plans that produce unacceptable damages for one or more sectors, regardless of benefits to other sectors. For example, IJC and LOSLR Study representatives reported that the economically most beneficial plan, particularly beneficial to hydropower, was rejected because it produced disproportionate losses to the environmental sector. The Board Room spreadsheet displays losses from selected plans, but it does not provide a record of the versions of those and other plans rejected as unacceptable, the progress made toward Pareto optimality through the SVM, the quantitative criteria for judgments of disproportionate loss, or changes in those criteria over time (though again see Leger et al., 2005). The IJC and the public would be informed by knowing the range of choice defined by both possibility frontier and loss criteria, as well as the record of tradeoff analysis made through SVM modeling. It is not clear from the SVM documentation which tradeoffs were made when various plans were screened, selected, or rejected. This information is crucial to create a transparent process. **The SVM plan evaluation process should document the range of efficient and minimally acceptable plans considered in order to facilitate informed decisions.**

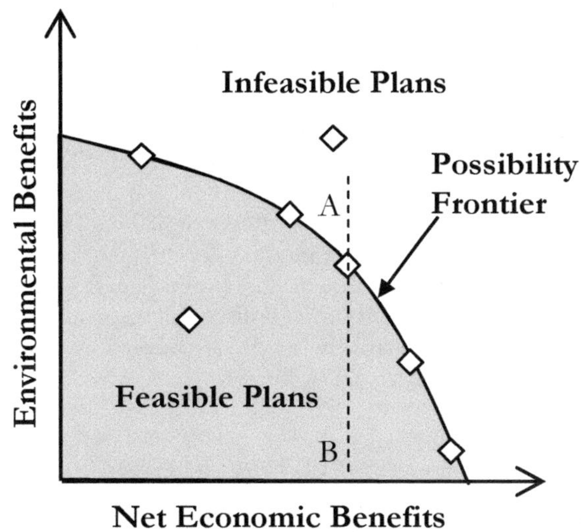

FIGURE 2-4 The possibility frontier for regulation plans with two objectives.

Comparable Treatment of Economic and Environmental Aspects of Performance Indicators

The LOSLR Study treats economic and environmental effects differently. While it is acceptable to forgo a full multi-attribute utility theory model (that is, not put economic and environmental indicators on common footing, whether via multi-attribute utility models or via monetizing environmental indicators), environmental and economic indicators should be treated in ways that enable proper integration of the two types of indicators in decision making. For example, while environmental performance indicators were subjected to an arbitrary "10% rule" to assess their significance, economic indicators were not. There is no statistical or practical justification offered for the decision to use the "10% rule" to express confidence in the significance of environmental performance indicator ratios.

The environmental indicator ratios are problematic. Although ratios may be reasonably well-behaved when used to evaluate plans that represent minor deviations from a baseline plan, they do not allow expression of variability and confidence limits to assess whether differences between plans are significant. In addition, the ratios were constructed in a variety of ways (sometimes aggregating before and sometimes after forming ratios, using moving averages for some indicators, etc.), which makes it difficult to compare dissimilar metrics across plans consistently, leading to inconsistent and unstable ordering of plans (SAR 3A discusses these issues nicely).

The Economics Advisory Committee raised a similar question about the comparability (i.e., fungibility) of economic variables. Economic indicators, although not a specific subject of this review, are a major aspect of the SVM, and the report of the Economics Advisory Committee (2004) was supplied to the Committee as a background document. The Committee shares the Economics Advisory Committee concerns about the treatment of economic factors in the SVM. It is not clear from SVM documents which of the Economics Advisory Committee recommendations have been implemented. For example, while the SVM has "net benefits" as an evaluation criterion, it does not follow standard procedures for evaluating the time streams of economic benefits associated with alternative plans. A generally accepted method for treating economic consequences over time is to use the present value of discounted benefits and costs over an economic time horizon. The present value of discounted stream of net benefits over T years (PVB) is defined as annual benefits (B) discounted by a rate that reflects economic time preference (r) (EAC, 2004, pp. 5-6):

$$PVB = \sum_{i=1}^{i=T} B_i/(1+r)^i$$

In the SVM application, the time horizon is 101 years. Values of PVB can be calculated for each 101-year sequence of inflows. Expected values of PVB could be estimated as the average value of PVB for multiple sequences of inflows. In this application, the 50,000 years of simulated flow sequences would provide about 500 possible sequences and 500 possible values of PVB. This series could also be used to estimate a probability distribution for PVB.

By holding external economic factors at present levels, the Study appears to have estimated only values of current benefits, B_0, under a range of possible annual sequences. This value does not capture effects of changes in economic factors and changes to physical and biological conditions over time.

The Study Board is aware of these issues (LOSLR, Year 3 Report, 2004, pp. 54-55). The coastal erosion performance indicator discounts future expenditures on shore protection, which vary under different hydroclimatic and water regulation scenarios (FEPS 7, pp. 23-7; and FEPS 14, p. 4). However, it is not clear how economic time preference has been addressed across different economic sectors and submodels in the SVM; and it is even less clear whether the present value of future environmental changes has been considered (e.g., as matters of intergenerational equity).

Because of problems with the way individual indicators were treated, and the different ways in which environmental and economic indicators were treated (e.g., the "10% rule"), the SVM output does not enable decision makers to comprehensively weigh trade-offs between environmental and economic indicators, as they will need to do in the absence of a fully formulated multi-attribute model.

Criterion 9: Balance of Scientific and Practical Professional Approaches

Evaluating the balance between scientific and practical professional approaches in the Shared Vision Model is contingent upon its performance under the other evaluation criteria, its constitutive models (FEPS, IERM, etc.), and its documentation. The constitutive models are discussed in chapters 3 and 4. This section focuses on the balance achieved within and among the other evaluation criteria. Each criterion has scientific and professional aspects, but the main scientific evaluation criteria employed in this review focus on empirical foundations, QA/QC, and treatment of error and uncertainty (Criteria 1, 2, and 3). Criteria 4 and 5 on study linkages and feedback, and spatial and temporal resolution, respectively, have strong scientific as well as practical professional roles in the LOSLR Study. Documentation, communication, and breadth of study scope criteria are common to both practical professional and scientific approaches. Two examples of the balance in the SVM are presented: one is drawn from the scientific criteria (Criterion 2) and the other from the practical professional side (Criterion 6).

Peer review is a standard scientific practice (Criterion 2). According to SVM materials, the SVM was to undergo two rounds of review: an intensive internal review and now this NRC review at the end of the study. The internal "peer" review is described in SVM documentation (BACK 6):

> The Shared Vision Planning (SVP) process provided perhaps the most detailed and intensive collegial review in that all the research and data was incorporated into the SVM. All the scientific and technical information had to be encoded as algorithms in the model, and each part of the model was validated independently.

Through the course of this NRC review, however, it became clear that the SVM was likely neither "validated independently" nor presented for external peer review in refereed publications. Peer review among practitioners and scientists in refereed publications would have given strong indication as to whether the SVM has struck the important balance between the scientific and practical professional approaches.

Documentation is standard practice for scientists and practitioners alike. It is used to explain methods, analyses, and treatment of error and uncertainty within and across different audiences. Problems with SVM documentation have been noted (see Criterion 6). These problems contribute to a lack of clarity and understanding about why certain decisions were made, such as the choice of regression analyses or rationale for using a single lake water elevation. These decisions, if better defended in stronger documentation, could have shed light on the balance between scientific and practical professional approaches in the SVM. However, this balance cannot be duly assessed without improved documentation.

Remedying the lack of documentation and quality assurance is necessary to assess the overall balance between scientific and practical profes-

sional approaches in the SVM, and in some of its specific modeling choices and applications.

Criterion 10: Identification of Future Study Needs

The LOSLR Study was initially planned to span five years. Now at the end of its 5-year life, the enormous amount of work and resources expended on the SVM development will be put to the best use over a multi-decadal time line. The history of the current single static plan, Plan 1958DD, demonstrates the need for flexibility and adaptability in water regulation in this dynamic watershed. However, future study needs were not explicitly identified in SVM review materials, and the LOSLR Study would benefit from a long-term plan that identifies future needs and potential changes. In support of the SVM being used over a multi-decadal time scale to assess changes in performance indicator responses to water level management, three needs are identified: revision and maintenance of the SVM, further environmental scenario analysis (e.g., climate change); and development of a systems dynamics approach to water regulation.

Revision and Maintenance of the SVM

SVM materials did not disclose a plan to revise, maintain, and update data collection and modeling activities. Like the first iteration in most studies undertaken to support important and controversial political decisions, current versions of the SVM are based on numerous assumptions and simplified representations of the real system. As noted in earlier sections of this chapter, the SVM has several deficiencies: existing documentation is incomplete, and there are several important gaps in the modeling structure and database. As a first step, these revisions need to be made. After the revisions are made and a new regulation plan is selected, the new plan will need on-going evaluation to determine how well it satisfies performance objectives under conditions of dynamic change in the Lake Ontario-St. Lawrence River basin. The history of change in Plan 1958D/Plan 1958DD shows that a single, static plan cannot be maintained over multiple decades, and ability is needed to adapt plans to changing conditions. **For the LOSLR Study to have a lasting effect, the SVM needs initial revision and then continued maintenance to keep current with changing conditions in the LOSLR basin.**

Further Environmental and Socioeconomic Scenario Analysis

Scenarios of long-term environmental and socioeconomic change also need a prospective approach that can be adapted with advances in scientific understanding of the region. Climate change is one example of dynamic human-environment relationships that has environmental and socioeconomic implications. Most recent research on climate change in the Great Lakes region indicates that the direction of possible effects on water levels and flows (up or down) is highly uncertain—as are the magnitude, timing, and spatial variability of change at regional and local scales. Although the potential impacts of climate change on the resources of Great Lakes basin have been extensively researched (Croley, 2003; Lofgren et al., 2002; Lofgren, 2003, 2004; Sousounis, 2002), and addressed to some degree in SVM scenarios of climate change, the need for further analysis of this and other long-term environmental processes may be anticipated. Similarly, as noted earlier, socioeconomic trends will likely have substantial relevance for future water level and flow regulation. **The LOSLR Study takes an important step by examining scenarios of climate change and water regulation, but these and other environmental and socioeconomic scenarios will likely change and require further evaluation.**

Toward a System Dynamics Approach

System dynamics simulation is a powerful tool for understanding complex feedback relationships among social and environmental components of a system (Simonovic, 1999; Simonovic and Rajasekaram, 2004; Sterman, 2000). The STELLA model is a system dynamics model as well as the modeling platform for the SVM. However, the STELLA model is not an immediate good fit for purposes of the SVM for several reasons. The value of system dynamics simulation lies in understanding the relationships within a complex system, and it is not typically used to predict the future behavior of a system. The STELLA model has capabilities that extend far beyond the simple water balance application for which it is being used in the SVM. The STELLA model, as developed and utilized in the Study has a very limited purpose: to calculate the response of the LOSLR hydraulic system to modification of rules governing release of water from Lake Ontario and the upper St. Lawrence River. The structure of the system is defined by the physical configuration of the network of rivers and lakes and the structures that regulate the flow of water through this system. This physical model of the hydraulic system is rigid. It is characterized by the same-state variables, and it does not add or subtract any elements or introduce any new interactions. For a given set of inputs (inflows to Lake Ontario, inflows from tributaries) the role of any model is to calculate flows and water levels at locations within the network where this information is required for other modeling purposes. Based on the structure of the SVM and its sub-models, this task can be performed by a spreadsheet using mass balance equations.

The STELLA model could have been the modeling engine for all sub-models in the SVM, but for reasons unexplained, it is not being used that way. The important feature of the hydraulic network in the STELLA model is that it remains static: whatever is happening in other subsystems has no impact on how the water is being distributed through the network over time. On the other hand, changes in the operating rules (new plans) do have an effect both on the spatial and temporal distribution of water flows and levels and on how other subsystems behave and respond. Although the hydraulic system can be modeled independently of other subsystems, which may be the purpose for using the STELLA model for the SVM, this rationale is not explained in the documentation.

Partial implementation of system dynamics simulation has the effect of underutilizing the STELLA model in the SVM applications. The SVM should include feedback loops to account for treatment of exogenous dynamic variables such as socio-economic and demographic factors, but it does not in its current design. This limitation was previously discussed under Criterion 4. Had the SVM included these feedback loops, the STELLA model would be a suitable application to simulate those feedbacks. **This partial implementation of the systems dynamics modeling approach (i.e., using the STELLA model and not including feedback loops) is a major conceptual limitation of the SVM as used in this study.**

SUMMARY

The LOSLR study is commended for its incorporation of broader water management goals for the LOSLR system and its commitment to public participation. Technically, however, its formulation and evaluation of alternative water regulation plans using the SVM modeling system are not likely to serve the intended purpose as well as they could. The SVM does not document its exploration and evaluation of the range of choice among regulation plan options and associated mitigation measures. Its deficiencies in documentation and quality assurance are uncharacteristic of scientific and practical professional approaches, and it does not lay out priorities for future modeling improvements and applications. As discussed in criteria 1 through 7, some important linkages between the SVM and supporting empirical research, sub-models, and exogenous variables are found wanting.

In spite of these criticisms, the Shared Vision Model approach represents an enormous investment in science to support decision making, and that investment should be captured to the maximum extent. Doing so requires an ongoing commitment to maintain and augment the databases and the models developed during the LOSLR studies in an adaptive management framework.

Experience with the Plan 1958DD plan indicates that it is unrealistic to think that a single plan can be adopted and then implemented for years into the future, given the dynamic physical and social context of the LOSLR system. Whether the issue is climate change, population growth, or new technologies for energy production, any water regulation plan for the LOSLR will be need to adapt to the dynamism of the Lake Ontario-St. Lawrence River basin. The SVM modeling system provides the kernel for an adaptive management system, but only if resources are invested in repairing its deficiencies and managing it for long-term use.

Specific recommendations to improve the overall appropriateness and sufficiency of the SVM to inform decision making include the following:

- The decisions to treat economic and environmental indicators differently should be justified in the SVM documentation or changed
- Incompatibilities in temporal and spatial scales between the STELLA

model and SVM sub-models could introduce untold error and need fuller exploration and explanation

- For the LOSLR Study to have a lasting effect, the SVM needs continued maintenance and updating to keep current with changing conditions in the LOSLR basin
- A systematic process is needed to document the unfolding range of choice among alternative regulation plans.
- Use of the STELLA model in the SVM represents a partial implementation of the systems dynamics modeling approach and is a major conceptual limitation of the SVM as used in the LOSLSR study.

REFERENCES

BACK 2. IJC (International Joint Commission). 1999. Plan of Study for Criteria Review in the Orders of Approval for Regulation of Lake Ontario-St. Lawrence River Levels and Flows. Available on-line at *http://www.ijc.org/php/publications/html/pos/pose.html.* Accessed September 2005.

BACK 6. Peer Review Triage Strategy (draft).doc.

Cardwell, H., B. Faber, and K. Spading. 2004. Reservoir Operating Plan Evaluation (ROPE) Study for the Mississippi Headwaters Collaborative Models for Planning in the Mississippi Headwaters. Available on-line at *http://www.iwr.usace.army.mil/iwr/svp/rope.htm.* Adapted from Conference Proceedings of the ASCE Environmental and Water Resources Conference, Salt Lake City, UT., June 28-July 1, 2004.

Cohon, J. L. 1978. Multiobjective Programming and Planning. New York: Academic Press.

Cohon, J. L., and D. H. Marks. 1975. A review and evaluation of multiobjective programming techniques. Water Resources Research 11(2):208-220.

Connor, J. L., L. Cartwright, and K. Stephenson. 2004. Collaborative water supply planning: A shared vision approach for the Rappahannock Basin in Virginia World Water and Environmental Resources Congress 2004. In Sehlke, G., D. F. Hayes, and D. K. Stevens (eds.) Critical Transitions in Water and Environmental Resources Management. ASCE Environmental and Water Resources Conference, Salt Lake City, UT, June 27–July 1, 2004.

Croley II, T. E. 2003. Great Lakes Climate Change Hydrologic Impact Assessment. IJC Lake Ontario–St. Lawrence River Regulation Study. NOAA Technical Memorandum GLERL-126, Great Lakes Environmental Research Laboratory, Ann Arbor, MI, September 2003.

Fan, Y., and D. Fay. 2001. Variation of St. Lawrence River Hourly Water Levels About the Quarter-Monthly Mean. Hydrology and Hydraulics Technical Working Group. Lake Ontario–St. Lawrence River Study, International Joint Commission.

Fan, Y., and D. Fay . 2002. Final Report on the Development of Empirical Relationships to Estimate Water Levels of the St. Lawrence River from Montreal to Trois Rivères. Great Lakes–St. Lawrence Regulation Office, Meteorological Service of Canada–Ontario Region, Environment Canada. March 28, 2002 (with corrections of October 2002).

Fan, Y., and D. Fay. 2003. Estimation of Lake Erie Outflows for Use in the Lake Ontario–St. Lawrence River Study. Great Lakes–St. Lawrence Regulation Office, Meteorological Service of Canada–Ontario Region, Environment Canada. December 20.

Goicochea, A., D.R. Hansen, and L. Duckstein. 1982. Multiobjective Decision Analysis with Engineering and Business Applications. New York: John Wiley & Sons.

Great Lakes Climate Change Hydrologic Impact Assessment I.J.C. Lake Ontario–St. Lawrence River Regulation Study. NOAA Technical Memorandum GLERL-126, September 2003.

Hagen, E., and J. Kiang. 2003. Interstate Commission on the Potomac River Basin Water Supply. Available on-line at *http://www.iwr.usace.army.mil/iwr/svp/icprbstudy.htm.* Accessed on. August 7, 2005.

IWR (Institute for Water Resources). 1995. *National Study of Water Management During Drought: The Report to the US Congress* IWR Report 94-NDS-12 (September).

Leger, W., F. Lupi, and W. Werick. 2005. Lake Ontario St. Lawrence River Study: Plan evaluation, rankings, and tradeoffs. Draft July 13, 2003. Copy on file at The National Academies, Water Science and Technology Board, 500 5th St., NW Washington, DC 20001.

Lofgren, B. M. 2003. Simulation of possible future effects of Greenhouse Warming on Great Lakes water supply using a regional climate model. Proceedings, 17th Conference on Hydrology, Long Beach, CA., February 8-13, 2003. Boston, MA: American Meteorological Society.

Lofgren B. M. 2004. Global warming effects on Great Lakes water: More precipitation but less water? In Proceedings of the 18th Conference on Hydrology, 84th Annual Meeting of the American Meteorological Society, Seattle, WA., January 11-15, 2004.

Lofgren, B. M., F. H. Quinn, A. H. Clites, R. A. Assel, A. J. Eberhardt, and C. L. Luukkonen. 2002. Evaluation of potential impacts on Great Lakes water resources based on climate scenarios of two GCMs. Journal of Great Lakes Research 28(4):537-554.

LOSLR Study (Lake Ontario-St. Lawrence River Study Board [LOSLR]). 2005. Options for Managing Lake Ontario and St. Lawrence River Water Levels & Flows: Final Report. July 1, 2005 Draft. Copy on file at The National Academies, Water Science and Technology Board, 500 5th St., NW Washington, DC 20001.

National Assessment Synthesis Team. 2000. Climate Change Impacts on the United States. A Report of the National Assessment Synthesis Team–US Global Change Research Program, 2000. Available on-line at h*ttp://www.usgcrp.gov/usgcrp/Library/nationalassessment/overview.htm.* Accessed September 2005.

NRC (National Research Council). 2004. Adaptive Management for Water Resources Planning. Washington, DC: The National Academies Press.

NRC. 2005. Valuing Ecosystem Services: Toward Better Environmental Decision-Making. Washington, DC: The National Academies Press.

Palmer, R. N. 1998. A history of shared vision modeling in the ACT-ACF comprehensive study: A modeler's perspective, presented at the 25th Annual Conference on Water Resources Planning and Management, Chicago, IL., June, 1998.

Palmer, R. N. 1999. Modeling Water Resources Opportunities, Challenges, and Trade-offs: The Use of Shared Vision Modeling for Negotiation and Conflict Resolution, Keynote Invited Presentation, presented at the 26th Annual National Conference, Water Resources Planning and Management Division of ASCE, Tempe, AZ., June, 1999.

Simonovic, S. P., and H. Fahmy. 1999. A new modeling approach for water resources policy analysis. Water Resources Research 35(1):295-304.

Simonovic, S. P., and V. Rajasekaram. 2004. Integrated Analyses of Canada's Water Resources: A System Dynamics Model. Canadian Water Resources Journal 29(4):223-250.

Sousounis, P. J. 2002. Special issue on Global Warming. Journal of Great Lakes Research 28(4).

Stephenson, K. 2003. The what and why of shared vision planning for water supply. A paper prepared for the panel session on Collaborative Water Supply Planning: A Shared Vision Approach for the Rappahannock River Basin, Universities Council on Water Resources, Water Security in the 21st Century Conference. Washington, DC. July 30, 2003. Available on-line at *http://www.iwr.usace.army.mil/iwr/svp/WaterSecurity03Text.pdf.*

Sterman, J. D. 2000. Business Dynamics–Systems Thinking and Modeling for a Complex World. New York: McGraw-Hill.

SVM 1. LOSLR Study Board. n.d. Evaluation Process: The Shared Vision Model. Provided for review as SVM document.

SVM 2. Shared Vision Model for the Design and Evaluation of Alternative Regulation of Lake Ontario Releases into the St. Lawrence River. Includes Toronto Board Room April 24.xls.

SVM 3. Shared Vision Model for the Design and Evaluation of Alternative Regulation of Lake Ontario Releases into the St. Lawrence River. Documentation Report. February 15, 2005.

U.S. Water Resources Council. 1983. Economic and Environmental Principles and Guidelines for Water and Related Land Resources Implementation Studies. Available on-line at *http://www.iwr.usace.army.mil/iwr/pdf/p&g.pdf.* 48 CFR 10249 (March 10).

Werick, W. 2000. The future of shared vision planning. In Hotchkiss, R. H. and M. Glade (eds.) Water Resources 2000: Building Partnerships. Joint Conference on Water Resource Engineering and Water Resources Planning and Management 2000, Minneapolis, MN., July 30–August 2, 2000.

Werick, W. 2005. Presentations on shared vision planning and modeling in the LOSLR Study to the NRC committee. Niagara-on-the-Lake, Ontario, CA. June 11-13, 2005.

Wescoat, J. L. Jr. 1987. The 'Practical Range of Choice' in water resources geography. Progress in Human Geography (1987): 41-59.

White, G. F. 1964. Choice of Adjustment to Floods. Research Paper No. 93. Chicago: University of Chicago, Department of Geography.

Chapter 3
Coastal Processes in the LOSLR Study: Flooding and Erosion Prediction System

The Flooding and Erosion Prediction System (FEPS) is a sub-model in the SVM that forecasts the physical and economic impacts of water regulation options on coastal erosion and flooding. FEPS models wave and erosion physical processes of public concern: intermittent flooding, erosion of barrier beaches, sediment transport, and the economic damage associated with coastal processes along the LOSLR shoreline.

The FEPS model and the field of coastal erosion and flood damage prediction remain partly in the domain of research and partly in engineering practice. Engineering design analysis usually aims to exceed a minimum factor of safety, whereas predictive engineering analysis forecasts failure conditions that are based upon and reveal failure processes. The FEPS model(s) are in the realm of predictive engineering analysis.

This chapter describes and evaluates the LOSLR Study documents about the FEPS model and its sub-models. The first section describes the FEPS model and its accomplishments; subsequent sections evaluate FEPS in terms of its scientific and engineering foundations, its ability to integrate and display key information, and overall sufficiency to be used to inform decision making about regulation plans.

The FEPS model analyzes three coastal performance indicators (PIs) that are included in the Shared Vision Model: [1]

1. Erosion PI – Lake Ontario/Upper St. Lawrence River
2. Flooding PI – Lake Ontario/Upper St. Lawrence River
3. Shore Protection PI – Lake Ontario/Upper St. Lawrence River

Three other coastal performance indicators were discussed in descriptive documents and contextual narratives, but they were not quantitatively modeled: Barrier Beaches and Dunes; Beach Access; and Sediment Budget.

Documents Presented for Review

The IJC presented the committee with several documents that represent the coastal processes work. The following documents were presented for review; the complete list of documents, including background documents and those distributed at committee meetings, are listed in Appendix B:

---

[1] Three related performance indicators modelled in the Lower St. Lawrence River with the RSPM model are not part of this review.

- ***FEPS 1*** – Zuzek, P.J., and R.B. Nairn. n.d. Automated Lakewide Flooding Predictions and Economic Damages on Lake Ontario.
- ***FEPS 2*** – Nairn, R.B., and P.J. Zuzek. n.d. Automated Lakewide Erosion Predictions and Economic Damages on Lake Ontario.

In addition, the Committee received fourteen background documents on the Coastal Performance Indicators (PIs), the Contextual Narratives associated with them, and oral presentations by FEPS and LOSLR staff at the open session meeting.

The condition of the materials submitted for review enabled an assessment of the general scientific validity and confidence in forecasting approach and results, but not modeling specifics (or model code). Thus, it is possible that some important details about FEPS have been omitted from this review.

DESCRIPTION AND ACCOMPLISHMENTS OF THE FEPS MODEL

The FEPS is a modular software system developed by Baird & Associates, and the overall modeling framework is fundamentally sound. It links bathymetric, topographic, and land use data in a GIS platform with engineering models intended to characterize the physical processes involved in coastal erosion and flooding, including wave generation through loss of shoreline, damage to structures, and shore protection costs (Figure 3-1). FEPS consists of several sub-models: a wave model (WAVAD), erosion model (COSMOS), and hazards models that estimate physical and economic damages of flooding and erosion. These models employ GIS tools and a relational database with results displayed in a user interface.

The modular structure of FEPS permits the replacement of individual sub-models with alternative models as they are developed. Modularity also can support alternative sub-models when needed to perform a multiple-model analysis. The FEPS model as a whole, and its engineering sub-models, are relatively new. It was originally developed for the Lake Michigan Potential Damages Study to assess hydroclimatic and water level impacts, and it was adapted for the Lake Ontario and upper St. Lawrence River system in this study.

Physical Modeling

FEPS uses two physical sub-models, WAVAD and COSMOS to model physical erosion and flooding processes, respectively. The numerical WAVAD model generates wave fields in a lake basin with varying winds and bathymetry (Blomgren et al., 1997; Resio and Perrie, 1989). The WAVAD model was used to hindcast wave fields in Lake Ontario using hourly wind field data from 50 climate stations to generate waves on a 3 kilometer grid in the Lake Ontario basin (Scott et al., 2004).

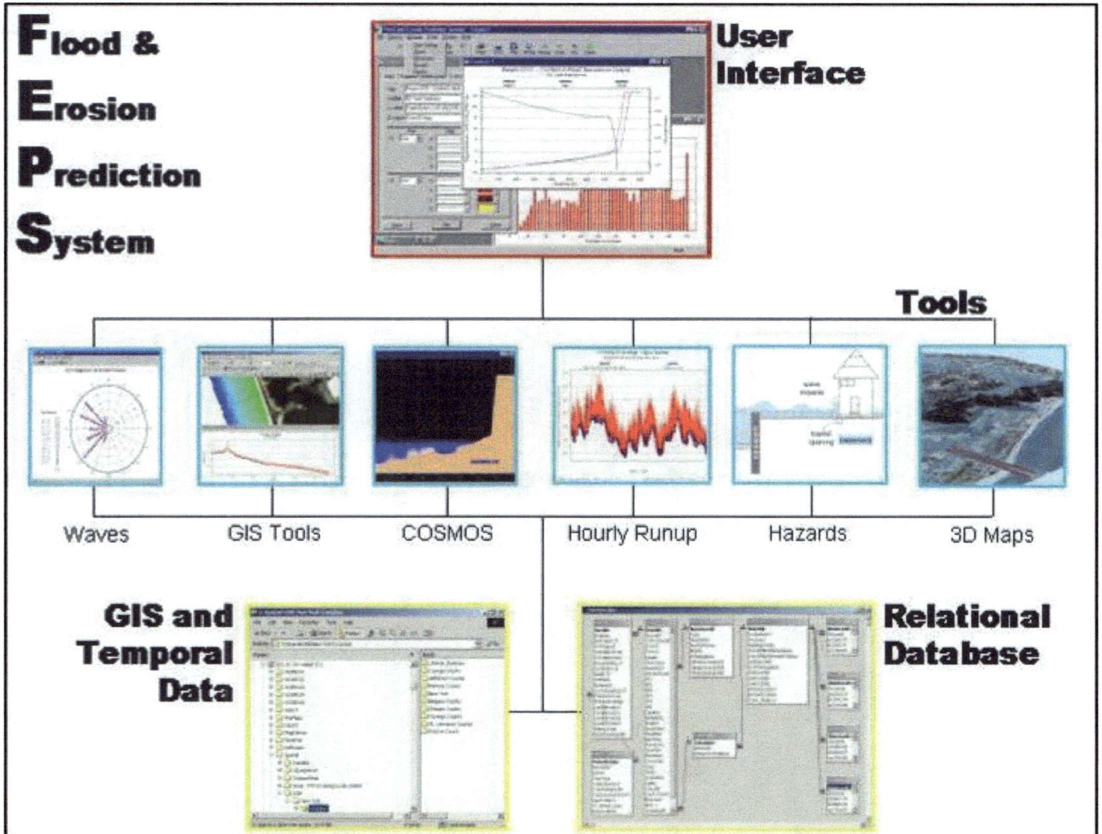

FIGURE 3-1 Structure of FEPS. SOURCE: Baird Software website (2005). Available on-line at *http://www.bairdsoftware.com/bairdsoftware/en_html/feps.html*. Accessed October 2005.

Hourly wave heights and directions were predicted at each grid cell for 1961 to 2000 CE (FEPS 2, p. 4). The period of record was extended to 101 years using a random selection of years from the historic record applied to the 1900 to 1960 period (FEPS 7, p. 9).

COSMOS is a process-based numerical model used to calculate coastal sediment transport and morphology (Southgate and Nairn, 1993; Nairn and Southgate, 1993). The synthetic hourly wave data generated by WAVAD for perimeter grid points along the Lake Ontario coastline were used as inputs to the COSMOS submodel. Full details about COSMOS were not available for review because it is a proprietary model. FEPS modelers noted that in the LOSLR application, the COSMOS model effectively erodes but does not aggrade due in part to Lake Ontario shore conditions and in part "to the inability of modeling sandy shore evolution over long time periods (i.e., 101 years)" (FEPS 10, pp. 1, 3; Zuzek, 2005).

The outputs of the WAVAD model were combined with an historical Average Annual Recession Rate (AARR) to predict coastal recession using regression equations calibrated by the COSMOS model. The documentation (e.g. FEPS 1) indicates that separate regression coefficients were developed for a site in water level increments of 0.25 m. The regression coefficients were treated as a family of equations to be capable of predicting bluff recession for other hydrographs that were developed as the study progressed. Lake-bed profile shape is treated as being similar across a county, as is near-shore wave energy. Regression equation predictions are calibrated to within 10 percent of the COSMOS model predictions in order to reduce and simplify the computational effort to meet the IJC requirements.

Flood and Erosion Hazards Modeling

Hazards modeling involved a combination of physical and economic analyses for erosion and flooding. This methodology was developed specifically for Lake Ontario.

Erosion Modeling

Erosion hazards were modeled as physical failures of shore protection structures and economic investment in future shore protection. Erosion of reaches with shore protection was modeled separately from the erosion of unprotected shorelines described above. It entailed three failure modes: overtopping of structures (based on county design data, lake level and wave height modeled with WAVAD); downcutting (or undermining estimated with the COSMOS model); and degradation of different types of structures (e.g., aging of Level 1 or Level 2 revetment) (FEPS 9).

Economic impacts of shoreline erosion were quantified based on the cost of designing and constructing shore protection to mitigate the erosion hazard over the 101-year simulation period. This damage avoidance approach assumes that accelerated shoreline erosion would lead shoreline owners to construct shore protection sooner and repair them more frequently, while plans with slower erosion rates would delay the cost of shore protection.

Flood Damage Modeling

Flood damages were modelled as physical and economic impacts on coastal buildings. An algorithm was developed to determine flood damage from inundation and damage from waves striking a building. The damages are quantified directly from the depth of inundation and the wave energy, respectively, using 4th order regression equations that relate the percentage of damage to flood elevation and to wave energy at the ground floor, respectively. The flooding algorithm does several things:

(1) extracts relevant property parcel data from the database on a parcel by parcel basis

(2) searches for lake elevation forecasts that exceed the main floor elevation of the structure on the parcel

(3) calculates the percentage damaged using the Federal Emergency Management Agency's depth-damage curves

(4) searches through water levels at each individual parcel and identifies lake levels that produce standing water between the lake and the house, based on ground floor elevations

(5) forecasts economic costs to the owners associated with these damages, by making assumptions about owner's responses to the hazard.

A similar process is applied for wave damage, but wave damages are computed using the "damage equation published by the Ontario Ministry of Natural Resources and Environment Canada (1981) based on recorded damages during the 1973 high lake level period" (FEPS 2, p. viii).

Three methods were used to estimate flood damages. The first method assumes that flood damages occur to homes with depreciation after each event but no repairs to the structure. The second method adds estimated costs for repair of homes and contents after each flood event. The third method allows for expenditure on structural mitigation of flood damages, set as a proportion of building value, which the FEPS study regards as the most realistic model of behavior (FEPS 1, pp. ix, x; FEPS 8, p. 2).

EVALUATION OF FEPS

This section assesses the degree to which the FEPS studies (A) reflect reasonable scientific methods, assumptions and supported findings; (B) integrate and display key information to evaluate tradeoffs for selecting among candidate regulation plans; and (C) are sufficient and appropriate for evaluating the regulation plan options and impacts of changes in water levels and flows.

EVALUATION OF SCIENTIFIC METHODS, ASSUMPTIONS, AND SUPPORTED FINDINGS

The first charge in the statement of task asks the degree to which the scientific foundations of the FEPS model(s) are sound. Three criteria were used to address the charge: empirical and theoretical foundations, quality assurance and quality control, and treatment of error and uncertainty.

Criterion 1: Empirical Foundations

The FEPS models combine analyses of winds, waves, coastal erosion, shore protection, physical flood damages, and flood protection costs. Detailed databases were compiled at the county, township, reach, and parcel scales for the models (FEPS 7, pp. 4-5). While some of the data-sets have a high resolution and long stable records, others consist of shorter-term records and in some cases proxy data (e.g., shore protection costs are used as a measure of erosion damages). This range of data quality poses challenges for FEPS to achieve its objective of making 101-year forecasts of coastal erosion and flooding damages under alternative water level and flow regulation plans.

The FEPS model is based on physical models and empirical equations. Some conditions of the models or the underlying science weaken the empirical foundation of FEPS. Since the FEPS model involves physical models and empirical equations, explicit treatment of the empirical basis of the models is expected. In particular, weaknesses in the data and the empirical bases of the models should be explicitly addressed. For example, FEPS documentation provided few details about the data quality or explanatory power of COSMOS model and the regression models. Data quality issues are mentioned for some topics, such as shore protection structures (FEPS 9, pp. 17ff), and previous research by Baird & Associates and others is mentioned but not included or extensively cited in review materials.

Some of the science underlying the FEPS work is fragmented, semi-empirical, or not independently repeatable. In cases with inadequate empirical data, expert judgment was used to replace or supplement the problematic data. The flood damage models lack data on flooding impacts and the associated economic damages for lake levels above 76.0 m (249.3 ft) on Lake Ontario because the lake has never reached these elevations. In this instance, the modelers used their "...professional judgment to evaluate the predicted flooding damages" and these predictions were deemed "reasonable" (FEPS 15, page 5). It is not clear how the judgment of reasonableness was determined, but some measure is needed to convey how reliability of this judgment.

The 2-Dimensional WAVAD model for Lake Ontario uses historic data (1956–1987) to generate hourly wind fields, and then hourly wave predictions on a 3 km-by-3 km grid for the period from 1961 to 2000. This 2-D model is important to the LOSLR Study because it is used in conjunction with the coarse quarter-month time step of the SVM with finer time steps. However, the documents reviewed do not indicate exactly how these data are used in the 101-year simulation period to calculate flood and erosion damages.

Regression Analyses

The FEPS work uses regression equations, in part, to predict shoreline erosion. There are good reasons to use regression for this type of work: regression is straightforward and inexpensive. Still, the applicability of regression analysis for FEPS has limitations that make it problematic for use in the FEPS analyses. One major limitation is the variable quality of FEPS data that would be used as a predictor for complex coastal processes over the next 101 years. The FEPS documentation provided for review did not allow for the linear logic, quality of fit, or error of the regression models to be ascertained. The high values of R^2 can only be interpreted with respect to historical data used in the regression analysis, and not the 101-year simulation period. The limitations of the regression equation-based modeling are not adequately discussed (see Box 3-1) in the FEPS materials, and implications of using regression analysis for FEPS applications need further explanation and defense.

BOX 3-1
Limitations of Regression Analysis for FEPS Applications

Regression analyses are inexpensive, available, and straightforward, but they may not be well suited for use in the coastal and erosion analyses in the FEPS model and sub-models. High order polynomials (4th and 5th order relationships), like those used in the FEPS models, are practical and commonly used in engineering practice. Although they can be used to approximate any smooth curve to a given accuracy by choosing a high enough degree, their accuracy and precision as forecast models cannot be assumed when data are scarce and not repeatable or when the predictive power of the regression equation has not been verified.

The practical advantages of high order polynomials are offset by a number of disadvantages (NIST, 2005). For example, polynomials are not suitable for describing asymptotic behavior where the curve approaches a straight line as the variable x gets larger in magnitude or for representing curves or data with sharp discontinuities in value or slope. Polynomials are sometimes considered to be of very limited use because of numerical stability problems.

In the coastal analysis, the regression equation is a logarithmic relationship scaled by the Average Annual Recession Rate (AAAR) in meters per year. The number of properties affected by waves is characterized by a fifth order polynomial (FEPS 8, p. 8). The damage to buildings is represented by two fourth-order polynomials derived from limited historic data. Damage to shore protection is characterized in terms of failure mode specific (undermining and overtopping) semi-empirical rules. The undermining failure mode equation is scaled by the

continues

BOX 3-1 Continued

AARR. The AARR is thus one of the most important parameters of coastal recession, defined as cumulative monthly recession in which:

$$Recession = (a \times \ln(E) + b) \times AARR$$

"Recession" is monthly cumulative recession and AARR is in m/year. This equation prompts five observations that require explanation:

a) Constant "a" is not dimensionless and is expected to have the dimensions of 1/Energy or $[kg^{-1}] . [m^{-2}] . [s^{2}]$. These units need explicit statement

b) If "**b**" is non-zero, then in the absence of any waves, the coastal recession will be: - **b x AARR**.

c) Any monthly Normal Wave Energy **E** less than exp**(-b/a)** would result in negative recession

d) The forecasted values of **'E'** are not independent of **AARR**, because AARR is dependent on the historic wave energy and the forecast of **'E'** is based on the same historic wave energy

e) The data used for fitting the regression for recession displays high variability (fig. 3.6) (FEPS 7). Figure 3.6 indicates very high variability of data used for the regression. The quality of the fit and the error do not appear to have been assessed (at least it has not been documented)

f) Goodness of fit criteria (r2) was calculated with logged data that artificially improves this value

g) Independent variables are collinear, which can make estimators **a** and **b** unstable.

The simple logarithmic relationship used to calculate coastal erosion has some curious characteristics that require explanation. The dimensional nature of constant "a" requires an explanation as constants in regression equations are normally dimensionless unless there is a more fundamental underlying relationship.

There is an almost direct relationship between parameters "a" and "b" of the form **"b" ≈ -9.5 x "a"**, over the full ranges of "a" and "b". A linear curve fit of the data reveled a relationship **"b" = -9.4919 x "a" + 0.0234**, with a correlation coefficient R^2 = 0.9992, an almost perfect fit. Thus for normal wave energy of approximately 13,000 J/m^2, the recession will be negative. While this relationship between "a" and "b" might be a coincidence, it should be explained.

The curve of recession as a function of normal wave energy necessarily becomes flatter as E increases. This means that as normal wave energy increases beyond 100,000 J/m^2 the slope of the line "$\delta R/\delta E$" (and neglecting the dimensional nature of "a" and treating it as dimensionless as in the study) described by the regression equation tends to zero. The result is that very large storms, with a wave energy of say 1,000,000 J/m^2 only make a 20% additional contribution to coastal recession above say 500,000 J/m^2. There is no way to ascertain the curve of recession because the calibration is restricted to about forty years of data that do not include such a large but plausible storm.

The decision to use regression analyses in the FEPS modeling apparatus needs detailed explanation and defense, including a discussion of the limitations and implications of using regressions for FEPS work.

FEPS Databases

The FEPS parcel database is impressive in its detail. It was reported to be complete for the USA part and approximately 75 percent complete for the Canadian part. The remaining 25 percent of parcels is a point of concern, as it has unknown implications on the final model output (this point was also raised in FEPS 4). Some of these issues are mentioned in the Contextual Narrative (FEPS 12) but not incorporated in the FEPS or SVM modeling. The database needs to be complete to ensure that the maximum range of scenarios is captured in the FEPS modeling effort.

The FEPS economic database is important in estimating flood impacts. In fact, an economic Performance Indicator was used in the SVM Board Room to evaluate operational alternatives. Documents were not provided on the economic database for this review, but according to information provided via oral presentations, the economic data are static. Considering the variation of real estate values and dynamics of the regional economy over the long simulation period, this assumption is unrealistic. Economic trends are discussed in the contextual narrative and a supplemental report, but they are not included in the modeling (FEPS12, pp. 3-5; and Christian J. Stewart Consulting, 2004).

The empirical damage equations are also based on very limited data sets. For example, the wave damage to buildings is based on as little as one data set from the 1973 high lake level period. FEPS has compiled extensive datasets that have long-term value, but FEPS documentation does not fully discuss empirical weaknesses or data quality issues that affect application and interpretation of these datasets in the Shared Vision Model. **Failure to specify the bounds of the empiricism of the model and its calibration, as well as the lack of a clear statement of what was done for conditions outside those empirical bounds, are errors that should be corrected.**

Criterion 2: Quality Assurance and Quality Control

Quality assurance was expected to have been rigorous for the FEPS erosion and flooding results. One issue of QA/QC in the FEPS work is whether regression analysis calibrated against a deterministic 2-dimensional physically-based model COSMOS is an appropriate approach, and if so, whether the calibration and regression analysis was carried out correctly. Quality assurance procedures are also needed to translate coastal recession, property and infrastructure damage, and economic disruption into economic costs. Since the FEPS modeling endeavor is an integral part of the SVM, quality assurance of the integration of FEPS into the SVM is required.

Transparency of the FEPS models is needed to convey the coastal impacts of regulation plan options to the scientific, public, and decision-making sectors. The COSMOS model is described as "extensively tested and validated" with numerous published articles in "peer reviewed journals attesting to the accuracy and robustness of the COSMOS erosion predictions for sandy coastlines" (FEPS 10, year, p. 4). Testing, validation and peer review are appropriate means of QA/QC. In the case of COSMOS, the documentation should have described the extensive testing and listed the peer-reviewed journal articles that attest to COSMOS accuracy and robustness. Other measures, too, can be taken to ensure transparency, such as engaging outside experts to use informed judgment to confirm scientific integrity in determining assumptions, selecting models and data, and adjusting analyses. These measures can improve transparency through formal methods for eliciting expert judgment (e.g., Morgan and Henrion, 1990).

One aspect of quality control is for scientific peers to carry out the same calculations and arrive at the same answer. This is standard practice in the scientific and engineering fields. For this type of validation to occur, the calculation models must be specifically described and the data made available so that the calculations are repeatable. It is common practice for a study of this magnitude and complexity to depend on rigorous independent scientific peer review of each component of the FEPS modeling. The independent review process may also suggest alternatives to the existing methods that could improve efficiency, costs, and time.

The FEPS documents mention that some aspects of quality assurance were undertaken (e.g., FEPS 14, 5), and they include minutes from independent technical reviews in 2002 (FEPS 18) and a discussion paper in 2005 (FEPS4), but these documents do not provide details on specific quality assurance procedures. **Overall, the quality assurance procedures for the FEPS work should be more fully specified and discussed in the review materials.**

Criterion 3: Treatment of Error and Uncertainty

Uncertainty and Sensitivity Analyses

The analysis of uncertainties associated with data, methods, and models for coastal processes and hazards is important. Broadly speaking, uncertainty in the FEPS models can be characterized as follows (Hartford and Baecher, 2004, pp. 127-134; cf. NRC, 2000, pp. 40-45):

- **Natural variability** is associated with the "inherent" randomness of natural processes, manifesting itself as variability over time for phenomena that take place at a single location (temporal variability), or as variability over space for phenomena that take place at different locations but at a single time (spatial variability)

- **Model uncertainty** reflects the inability of a model or design technique to precisely represent a system's true physical behavior, or the inability to identify the best model, or a model that may be changing in time in poorly known ways (e.g., flood-frequency curve changes because of changing watershed)
- **Parameter uncertainties** result from an inability to accurately assess parameter values from test or calibration data, due to limited numbers of observations, and the statistical imprecision attendant thereto
- **Data uncertainties,** including (i) measurement errors, (ii) inconsistency and in homogeneity of data, and (iii) data handling and transcription errors, and (iv) inadequate representativeness of data sample due to time and space limitations
- **Decision model uncertainty** is also important in contexts such as the FEPS and Shared Vision Model, where decision-making objectives change over time (NRC, 2000, p. 44).

The methods used to deal with data uncertainty including gaps in the data should be available and their use justified. In particular, assumptions that gaps in the data are insignificant or of limited significance overall are unacceptable, and specific measures to address those gaps in the data would normally be required. The degree of resolution (error bounds) of all data should be specified and propagated through the modelling endeavour. Against this background, procedures are required to explain how the processes of interpolation and extrapolation have been embodied in the data analysis and modeling process.

Sensitivity studies involve determining the change in response of a model to changes in individual model parameter distributions or changes in the model. Sensitivity analysis is useful but not sufficient for characterizing model and parameter uncertainties. Thus, the output is ideally in the form of a probability distribution which specifies the likelihood of each possible result across the full range of possible results (Hartford and Baecher, 2004, p. 20). Probability distributions of parameters to which the analysis is sensitive are identified and their validity and accuracy verified to the extent that is realistic. Changes are made to probability distributions where justified.

Uncertainty and Sensitivity Analyses in the FEPS Model

The FEPS documents were expected to present a comprehensive theoretical and analytical treatment of the uncertainties in the models, parameters, and data. FEPS performance indicator summaries mention uncertainty (e.g., FEPS 4, p. 5), and the 2002 technical review recommends treatment of data and modeling uncertainties (FEPS 18, pp. 8, 12), but important details are not reported. For example, failure of shore protection structures is amenable to analysis using structural reliability methods (USACE, 2003, Voortman et al., 2002). The results of such an analysis for shore protection structures would yield a theoretical fragility curve (Figure 3-2, as adapted from NRC, 2000, p. 70). **Model uncertainty should have been dealt with explicitly, as should the propagation of uncertain data through uncertain models.**

The uncertainty in the logarithmic regression equation is the most problematic aspect of the FEPS study because of its capacity to introduce and propagate error. All downstream modeling considerations are dependent, in part, on the logarithmic regression output. This dependence places critical importance on those output values being accurate; otherwise error will be propagated through the downstream models. Another problem stems from the use of the synthetic data set of monthly normal wave energy calculated by WAVAD and monthly recession calculated by COSMOS. The fit for the recession regression equation uses these data. Using synthetic data introduces unnecessary error (E1) on recession outputs, which subsequently are used as inputs for the regression analysis. Regression analysis produces its own error (E2) resulting in the aggregate error of E1+E2. The first type of error (E1) could presumably have been avoided by skipping the COSMOS process and using the observed recession data instead.

The uncertainty in the data used to derive the logarithmic relationship exhibits a great deal of scatter (Figure 3.6 of FEPS 7, p. 11, reproduced below, Figure 3-3). The horizontal axis extends to 250,000 J/m^2 and covers the range of the average yearly normal wave energy. However, the WAVAD model predicts January wave energy in excess of 600,000 J/m^2. The rate of change of the predicted recession decreases rapidly as wave energy increases. This implies that recession during severe winter storms will be similar to that for significantly lesser storms. From the materials presented for review, it was not possible to determine whether this feature of the model is reasonable. Thus calibrating the recession regression equations

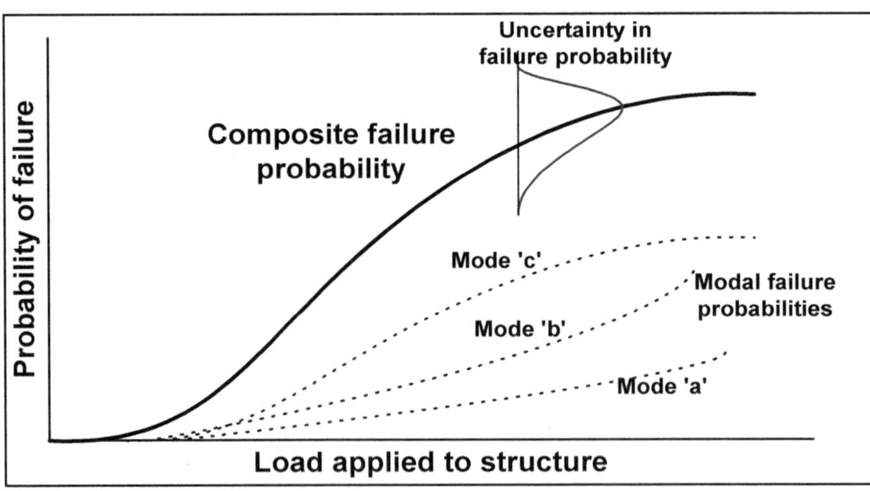

FIGURE 3-2 Form of result of structural reliability analysis of shore protection structure.

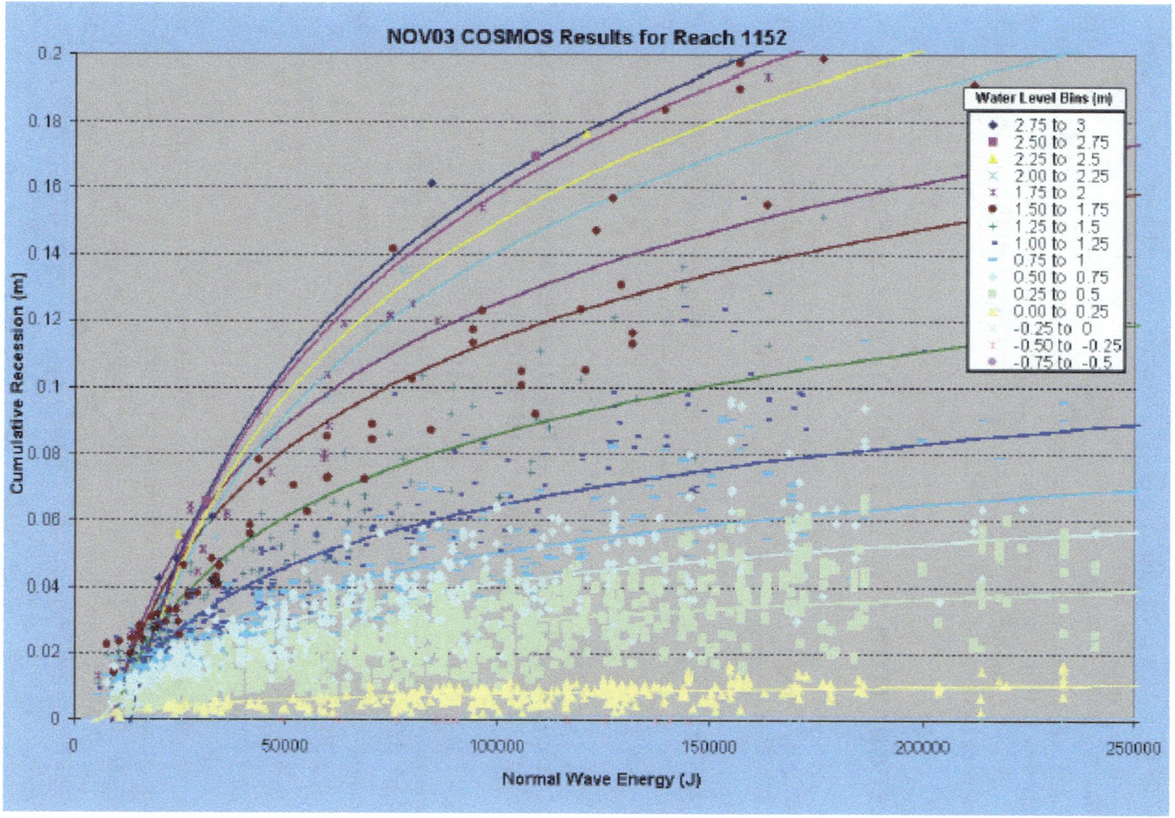

FIGURE 3-3 Cosmos modeling of cumulative recession (FEPS 7, p. 11).

to within less than 10% of the COSMOS model equations has little effect because the scatter around the COSMOS regression equations is greater than 10 percent. The uncertainty in parameters "a" and "b" must be significant, such that it may not be possible to make a reasonable distinction between the recession rates associated with different plans. As noted above, the problem of calibrating the model for conditions beyond the bounds of the empirical data introduces additional uncertainty that renders the forecasts questionable. **The lack of specific treatment of uncertainties in the coastal recession regression model, especially parameters "a" and "b," make objective discrimination between the coastal impacts of regulation plan options impossible.**

INTEGRATION AND DISPLAY OF KEY INFORMATION

The second charge in the statement of task focuses on the degree to which the FEPS models integrate and display coastal erosion and flooding information needed for a comprehensive evaluation and understanding of the tradeoffs for selecting among candidate regulation plans. Four criteria were used to evaluate this charge: linkages and feedbacks among related studies and models; treatment of spatial and temporal resolution and scale; documentation; and scientific communication of flooding and erosion outputs.

Criterion 4: Linkages and Feedbacks Among Related Studies and Models

The FEPS model is one of several sub-models of the SVM. Linkages between FEPS and the other LOSLR models, particularly the integrated ecological response model and the SVM, are considered internal linkages. Exogenous linkages are those between FEPS and external variables, such as economic and demographic values. These two types of model and study linkages are discussed in this section.

Internal Linkages

The structure of the SVM and its sub-models presents an opportunity for holistic understanding of the Lake Ontario-St. Lawrence River system, as independent processes are modeled independently and then brought together for a single purpose in the SVM. Despite connections between bathymetry and wetlands processes, the FEPS model output was not connected to the integrated ecological response model (IERM) in the SVM. This is a missed opportunity to advance the science and better understand intricate relationships in the LOSLR system, as the FEPS model output would have informed the wetlands scientists with needed information about wetland bathymetry and thus the areal extent of wetland vegetation and species at risk habitats. Furthermore, linkages between FEPS and the

IERM could have simulated flooding and erosion effects on environmental variables. Similarly, FEPS documentation does not indicate how erosion analysis in Lake Ontario and the upper St. Lawrence River is related to or comparable with downstream analysis in the Lower St. Lawrence.

As discussed in the FEPS documentation, the model does not provide for sandy shore processes (see FEPS 7, 10 and 13). This omission limits the ability to assay flooding and erosion of barrier beach wetlands in the Shared Vision Model and Integrated Ecological Response Model (see Chapter 4). It also misses an opportunity to represent inherent feedbacks among erosion, vegetation, resistance to erosion, and water regulation plans. A regulation plan can affect erosion in ways that result in different types of vegetation and that in turn cause a change in erosion resistance. The FEPS analyses do not incorporate these dynamic feedback relationships. The lack of beach accretion in the FEPS model may have implications for other areas, uses, and values of Lake Ontario shore areas.

To their credit, the FEPS performance indicator summary and contextual narrative underscore some of these risks and mention separate analyses undertaken to address them (FEPS 10 and 12).

Exogenous Linkages

FEPS treats exogenous processes including coastal land use and economic development as static, which results in the same weaknesses as in the Shared Vision Model (noted in Chapter 2). Trends in coastal land use, economics, and hazards mitigation are discussed in a contextual narrative but not incorporated in the model (FEPS 12; Christian J. Stewart Consulting, 2004). Since coastal erosion and flood losses are just two dimensions of a dynamic and interdependent environmental system, the FEPS model should at least identify and ultimately incorporate causal relationships with the other dimensions of the system. **The effects of coastal erosion and flooding should be incorporated as inputs to the environmental and economic models as well as the Shared Vision Model.**

Also, the FEPS modeling currently does not include an analysis of the potential effects of future water regulation deviations, analogous to the deviations that have occurred under Plan 1958D, on flooding and erosion damages under alternative hydroclimatic scenarios and water regulation plans. The FEPS model includes explicit forward linkages but limited feed-back linkages among its sub-models and no linkages with associated ecological or other socioeconomic models. Along with alternations to the shoreline by riparian owners in response to coastal erosion and flooding that are inconsistent with the modeling assumptions, deviations from the selected plan, represent considerable uncertainties that should be incorporated in the analysis.

Criterion 5: Treatment of Spatial and Temporal Resolution and Scaling

The range of temporal and spatial scales in the FEPS models is enormous. The temporal and spatial scales of inferences derived from the FEPS modeling endeavor are not currently explained in the FEPS documentation. Linking hourly wave data with quarter-monthly lake levels introduces issues of spatial and temporal resolution at the outset of the analysis. **Questions pertaining to the resolution of the flooding and erosion forecasts should be addressed comprehensively and objectively in the documentation to avoid misinterpretation.**

The FEPS modeling endeavor requires specific scales to be adopted for each of the sub-modeling efforts, making it necessary to scale the results of the analysis in both space and time. In Figure 3-4, "Time" reflects the temporal extent over which the erosion/damage parameter and planning horizon occur. "Spatial dimension" reflects the geographical extent and lake level fluctuation of those erosion/damage parameters and planning horizons. The FEPS process would be expected to illustrate where in this space-time context data are collected, processes modeled, and inferences made. The spatial and temporal resolution of elements in the FEPS model and sub models are illustrated conceptually in Figure 3.4. Describing the FEPS work in such a context would provide a basis for framing consultations, discussions, and debate around issues of modeling resolution and the sphere of applicability of FEPS results.

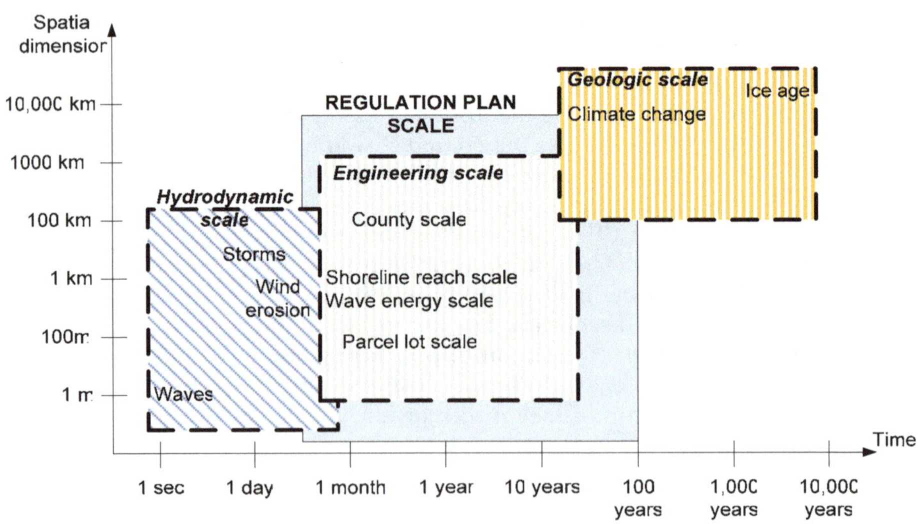

FIGURE 3-4 Spatial and temporal scales.

The SVM generates a 101-year simulation horizon in a series of single water elevations with a quarter-monthly time step (see Chapter 2) that is used in the FEPS algorithm. Both the single value (which is a spatially averaged value) and the quarter-month time step are too coarse for purposes of the SVM (see Chapter 2) and some environmental considerations (see Chapter 4). Similarly, the FEPS models are highly sensitive to actual lake elevation at the time of a storm event. In fact, the important parameter for the erosion model is the lake level at the time of the storm, not an averaged value over a 1-week period. Connections between the SVM and FEPS models appear to give inadequate treatment to the spatial and temporal variation of lake water levels within the time step important for proper evaluation of flood damages. The compatibility of data on different scales and their use to simulate future conditions raises a prominent concern. With reference to Figure 3-4, for example, the spatial and temporal scales of the economic damages are apparently restricted to:

- 1 km reaches of coastline ± an order of magnitude (FEPS 8, p. 20; FEPS 9, p. 25)
- 1 year (for damages scaled to the Average Annual Recession Rate; and for erosion damages defined by the year of investment in shore protection, FEPS 7, p 21).

The model documentation appropriately addresses issues of resolution. In its discussion, it identifies some limitations to the FEPS model. It specifically notes the lack of parcel-level detail that precludes more than an order of magnitude forecast at the reach scale (FEPS 8, p. 20), which means that FEPS cannot capture dramatic changes in local shoreline conditions that can occur during individual storms. The FEPS model, however, would be most useful under such conditions when it could convey predicted responses to interested parties, such as riparian and shoreline owners, the IJC, or others in the political process.

Relationships among spatial dimension, analytical effort, and the accuracy and precision of results are represented schematically in Figure 3-5. From the information available, it appears that the FEPS modeling endeavor is at the meso-scale spatially (FEPS 7, p. 3). The mathematical scaling in space and time of the "Families of Equations for Regional Zones" (FEq_i), the "Local Wave Energy (We_i)," and the Annual Average Recession Rate (AAAR) measured over each 1 km reach is not specified. The FEPS modeling endeavor is considered to be of low- to medium-analytical rigor, due in part to the regression equations and in part because there are at least two higher levels of analysis available. The highest level of rigor is structural reliability modeling on a parcel lot-by-parcel lot basis using detailed wave and hydraulic models. The next highest level of rigor is multiple COSMOS-type modeling on a 1-km reach basis over the entire lake and river.

In addition, spatial or temporal intra-annual autocorrelation may occur. The annual recession value is calculated as the sum of individual recession events that occur during the year. For example, the performances of two sections of a shoreline reach may exhibit a long wave length of correlation in space, and thus exhibit

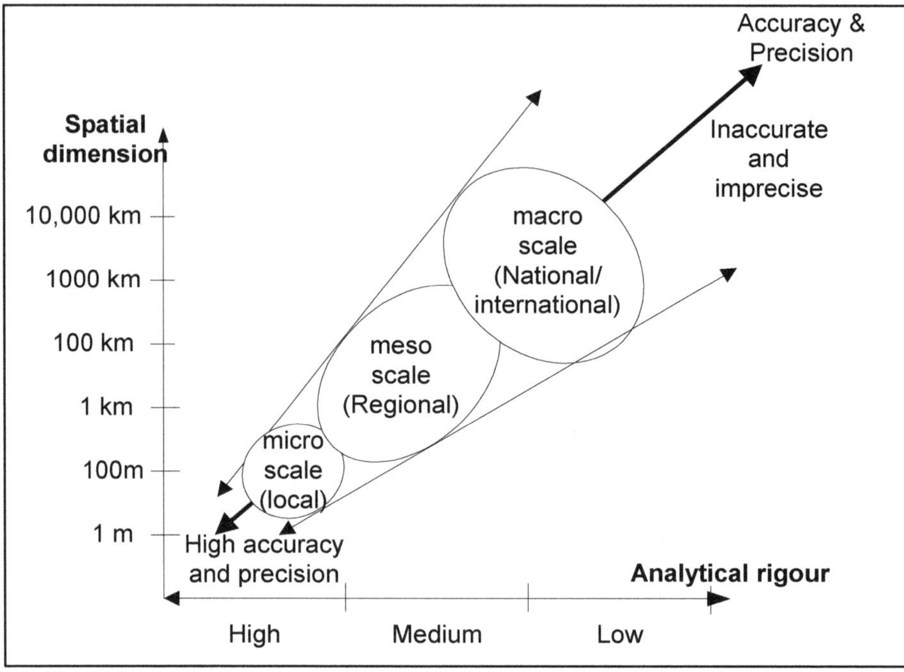

FIGURE 3-5 Spatial scale, analytical rigor, and accuracy and precision of results.

similar factors of safety against erosion. The FEPS review documents do not discuss these issues of spatial and temporal autocorrelation, or the effects of different storm sequences on shoreline evolution.

The treatment of spatial and temporal scaling requires considerable additional effort to enable a comprehensive evaluation and understanding of flooding and erosion hazards associated with regulation plan options.

Criterion 6: Thorough Documentation

Documentation needs to sufficiently inform a broad spectrum of users, especially for large, complex studies. It is an additional challenge to make this documentation navigable and interpretable to the full spectrum of users. A "map" of the documentation would provide a useful means of presenting studies of this nature. Documentation should completely describe both "what was done" and the justification for "why it was done the way it was done." The documentation should permit the scientific community to review and if necessary reproduce all calculations. This means that all data must be provided and all models completely specified.

Documentation insufficiencies are noted throughout this chapter. Notable documentation concerns relate to:

- Description of the limitations of the empirical foundations for FEPS models
- Treatment of uncertainty and error
- Quality control and quality assurance of model results
- Inferences based on professional judgments
- Robust justification for the way that coastal erosion and flood damage calculates the various inputs to the SVM.

There is no final report for FEPS, and several of the individual items of documentation are clearly interim documents describing modeling intent as opposed to finalized details of the modeling effort. For example, FEPS 7 states "Predictive equations from the SVM are being developed based on the total monthly recession predicted by COSMOS" (p. 10). That the documentation would be highly conditional and written in the language of uncertainty would be an expectation for the FEPS process. The FEPS documentation should ultimately also be suitable for publication in the broader body of established scientific domains of coastal erosion, hydrodynamics, soil mechanics, structural reliability, economics, long-term policy and planning.

Although superior to documentation of the SVM, FEPS documentation presented and reviewed here is deemed inadequate for external scientific review, and for the judgments and conclusions to be defensible in public decision-making.

Criterion 7: Effective Scientific Communication

Results from the FEPS study can be interpreted in several ways over different spatial and temporal scales and warrant particular attention to communicating uncertainties at various scales. **Protocols should be established to ensure that those relying on the results of the FEPS studies understand the strengths and limitations of the modeling endeavor, and the implications of interpreting the outputs in different ways.**

Forecasting necessarily demands a tradeoff between the "security" of the forecast and the "informativeness" of the forecast. "Security" of the forecast pertains to the extent to which the forecast can be accepted in advance of the actual outcome, whereas "informativeness" pertains to the extent to which the forecast is genuinely insightful, providing definitive information. "Security" of forecasts increases as the forecasts become more general whereas "informativeness" increases as the forecasts become more specific. "Security" often comes at the expense of "informativeness" (Rescher, 1998). The FEPS study documentation does not discuss the nature of such forecasting tradeoffs, or the final balance that it took to those tradeoffs.

The process of communicating the accuracy, precision, reproducibility, and reliability of coastal erosion and flood damage prediction is extremely difficult, and considerable effort is needed to ensure that the results are not readily misinterpreted. The problem of properly representing risk and uncertainty in coastal erosion and flood damage prediction in a public decision process should be addressed. Forecasting can be difficult over the short to medium term, with the result that decision-makers and users of model results often form their own degree of belief in the credibility of the results. **The communication of the tradeoff between the "security" and the "informativeness" of the forecast should be explicit and available to users to make an informed judgment as to model usefulness.**

The results of the FEPS modeling can also be expected to be interpreted and used by owners of individual parcels of property, insurance companies, municipalities, courts at all levels, governments of bordering states and provinces, and the two national governments, as well as by international NGO's and other parties with no direct interest in the regulation of Lake Ontario and the St. Lawrence River. Accordingly, the documentation and scientific communication should be prepared to meet the challenges posed by such a broad spectrum of users in such a way that the integrity of the studies is preserved. **Effective scientific communication of FEPS model output requires a more thorough discussion of uncertainties associated with the flooding and erosion forecasts to enable a broad public understanding of tradeoffs among regulation alternatives.**

OVERALL APPROPRIATENESS AND SUFFICIENCY TO INFORM REGULATION PLAN OPTIONS

The third charge of the statement of task is aimed at determining whether the IJC documents presented for review are appropriate and sufficient for use in selecting a water regulation plan option and determining impacts of changes in water levels. Three evaluation criteria are used to address this charge: study breadth, balance between scientific and practical inquiry, and the identification of future study needs.

Criterion 8: Breadth of Study Scope

FEPS encompasses a wide range of physical analysis and associated risk analysis for people and property. For example, FEPS analyzes poorly understood processes over a range of physical conditions (albeit not the full range) that could exist over the period of analysis and geographic scope of the Lake Ontario and upper St Lawrence River coastal zone. Under such circumstances, a heuristic approach would provide an effective framework for coastal erosion and flood damage prediction.

In other ways, the scope of the FEPS study remains narrow. Erosion hazards for environmental systems, public infrastructure, and wider economic activities are omitted from the model and only briefly discussed in conceptual narratives. The

FEPS focus on generating data that will be used in a standard deterministic economic cost comparisons may limit broader scientific understanding of the bases for water level regulation decisions.

Since the results of the FEPS analysis are utilized in the SVM and in public consultation, the FEPS documents should describe the design of interfaces between FEPS, other modeling endeavors, and the public consultation process to ensure that the modeling efforts are in harmony. Typically, these interfaces would include dynamic feedback from other models to permit comprehensive modeling of the system. **The FEPS study takes a broad approach to erosion and flood hazards of water level regulation of Lake Ontario and Upper St. Lawrence River, but greater breadth is needed for a comprehensive, scientific analysis of the erosion and flood hazards associated with water-level regulation in the LOSLR.**

Criterion 9: Balance Between Scientific and Practical Professional Approaches

The relationships between scientific inference and informed professional judgment should be elucidated. As noted above, protocols for qualifying experts and data should be defined and adherence to these protocols demonstrated. In the case of FEPS, the scope of the study and documentation should also indicate how precise and accurate the predictions are expected to be. It should demonstrate that each of the sub-models (wave, coastal erosion, failure of coastal protection structures, flooding and damage models) are representative of actual conditions. Different models will combine these considerations in different ways leading to different results. When modelling complex physical processes that are poorly understood such as coastal erosion and damages, the simpler the model the greater the difficulty in demonstrating the representativeness of complex processes (Pilkey and Cooper, 2004). Thus, modellers are challenged to demonstrate that (1) simplified erosion hazards models are sufficiently representative of the actual process and (2) model results lead to the same conclusion as detailed erosion hazards modelling.

Relationships between scientific research and practical scientific inquiry into matters of erosion, flooding and physical damage associated with flooding have become better defined in recent years with studies of this type and size becoming more common. The"EUrosion" project, (EUrosion, 2004a) is an example of a study that shows the understanding of coastal processes is still largely fragmented and semi-empirical (EUrosion, 2004b). The net result of such fragmentation, which is common to other developing sciences, is that different theories based on different concepts, assumptions and approaches result in different models with different levels of compatibility.

Balancing scientific and practical inquiry in FEPS should include specific discussion of:

1. The methods used to compensate for gaps in the physical data

2. The use of expert opinion throughout the FEPS modeling endeavor in relation to scientific norms
3. How the current weaknesses in the FEPS model might be improved through future research, testing, and regulatory adjustments, all of which would help refine the FEPS modeling process dynamically into the future.

The balance between scientific and dimensions of pragmatic inquiry is compromised by the impression of certainty that is conveyed throughout the documents. For example, the Erosion Performance Indicator document (FEPS 7) makes no reference to the uncertain state of scientific knowledge in coastal erosion and flood damage prediction, and the PI summary document mentions it only briefly (FEPS 14).

The FEPS model aims to provide coastal erosion and flood damage data for a deterministic economic analysis, and it appears that these issues of practicality dominated the design of the model. However, FEPS was not specifically developed for regulatory policy analysis, and it was first conceived for the Lake Michigan flood damages analysis. As a result, the model is limited in the damages it examines. The FEPS model considers flood and erosion damages to structures and shorelines but not to infrastructure, agriculture, or environment within an integrated approach, as has been the case in previous studies carried out for the IJC (2000) and others (e.g. Simonovic and Carson (2003)). Although pragmatic limits are necessary in studies of limited duration and funding, they should be explicitly mentioned along with their potential implications for decision-making.

FEPS attempts to balance scientific and professional methods of forecasting flooding and erosion hazards, but it does not describe the balance to be achieved or the extent to which that balance objective was achieved; and it does not sufficiently convey the challenges associated with this aim or the limitations of its approach

Criterion 10: Identification of Future Study Needs

The FEPS databases formed during the LOSLR Study have value and provide a useful foundation for the future work. Integration of the property database, wave model, and damage calculation model while incomplete is an important step. Predictive capability would be expected to improve over time, and there is opportunity to refine the FEPS modeling process dynamically over time.

A long-term monitoring program would help assess the reliability of the FEPS model predictions using water regulation outcomes. **Such a long-term monitoring program is required, and would address three specific needs:**

- Monitoring actual erosion and flood damage and improving the models using reliable data.
  - Given that the FEPS model and its sub-models have not been demonstrated as being capable of precisely and accurately predicting coastal erosion and flood damages over the 101-year simulated planning horizon, a feed-

back process that updates the forecasts at regular intervals of 5 to 10 years would be expected

- A dynamic process to revise FEPS models
  - In the case of coastal erosion and flood damage prediction, future studies should be identified to analyze the water regulation plan that is adopted, including any deviations, as well as changes in coastal land use, land value, shoreline protection technologies, and other flood damage mitigation measures
- Promotion of the advancement of scientific research on coastal erosion, flooding, and damage reduction in the Lake Ontario and St.Lawrence River basin
  - It is clear from this review and complementary studies elsewhere that forecasting coastal erosion and flood damage is in its infancy, that the scientific knowledge is fragmented and largely incomplete, and that additional research in the lower Great Lakes is necessary.

SUMMARY

The overall modeling framework for the Flooding and Erosion Predication System is impressive in many respects. It includes detailed parcel and reach analysis, and undertakes a lakewide analysis not previously attempted.

From a scientific perspective, and due in significant part to the lack of documentation, the reliance on regression equations that are poorly conditioned and inadequately justified, and the lack of formal treatment of uncertainty, the FEPS study as presented is insufficient for the purpose of enabling a comprehensive evaluation of tradeoffs among regulation plans. From the same perspective, the scientific robustness of the FEPS models remains to be demonstrated.

It is clear from the documentation that the regression analysis approach to modeling coastal erosion and forecasting flood damages is considered by the FEPS authors to be less than ideal. But this less than ideal approach was adopted because, "it is not practical for a parcel level application due to the geographic scope of the study, large data requirements, the manual nature of the desktop approach and the time required to generate answers. The study managers required an automated approach that produced robust and defensible results for the erosion impact assessment," (FEPS 1, Nairn and Zuzek, p. 7). This review concurs that the regression analysis approach in the FEPS application is less than ideal, but considers that computational techniques are available that can be applied more extensively at a parcel lot level. The mathematical properties, limitations and implications of the regression equations should be clearly explained to the decision-makers. In particular, the analysts should demonstrate that the regression equations have a physical basis and that the potential problems of numerical instability have been addressed.

In light of the multiple scales and models used, a comprehensive analysis of uncertainties is needed. Further, multiple model analysis of the coastal erosion process would be a significant improvement over single model analysis.

One issue that remains, and which cannot be resolved given the design of the FEPS study, is the extent to which the results of the analysis as carried out would differ from those that would be obtained through application of best available science over the entire system. A second issue that remains and which may never be fully resolved is whether these differences matter in practical terms.

Against this background, the degrees of robustness and defensibility of the FEPS studies reviewed here have not been demonstrated or documented in a comprehensive way. The lack of explicit treatment of risk and uncertainty limits confidence in FEPS results for use in informing decision making, although, unfortunately, FEPS documentation and presentation may convey an impression of accuracy and precision in the data, models and interactions within and external to the system. In addition, the failure to identify those risks that society cannot afford to take, and those risks that society cannot afford to take too often, render the study incomplete.

REFERENCES

Blomgren, S., H. Hanson, and M. Larson. 1997. Modeling of wave generation and transformation. Scientific report, Appendix A: Documentation. Subproject 4: Nearshore and coastline processes, 1 year progress report, Baltic Sea System Study

Christian J. Stewart Consulting. 2004. A Summary of Existing Land Use, Land Use Trends and Land Use Management Policies along the Lake Ontario-St. Lawrence River Shoreline: Implications for Future Water Level Management. Prepared for the Coastal Technical Working Group, International Joint Commission.

EUrosion Project. 2004a. Living with coastal erosion in Europe: Sediment and Space for Sustainability *PART I*—Major findings and Policy Recommendations of the EUROSION project Service contract B4-3301/2001/329175/MAR/B3, Coastal erosion—Evaluation of the need for action. Directorate General Environment, European Commission.

EUrosion Project. 2004b. Living with coastal erosion in Europe: Sediment and Space for Sustainability. PART IV—A guide to coastal erosion management practices in Europe: Lessons Learned Service contract B4-3301/2001/329175/MAR/B3. Coastal Erosion—Evaluation of the need for action. Directorate General Environment, European Commission.

FEPS 1. Zuzek, P. J., and R. B. Nairn. n.d. Automated Lake wide Flooding Predictions and Economic Damages on Lake Ontario.

FEPS 2. Nairn, R. B., and P. J. Zuzek. n.d. Automated Lake wide Erosion Predictions and Economic Damages on Lake Ontario .

FEPS 3. FEPS Model.

FEPS 4. Sciremammano, F., and S. Renzetti. 2005. FEPS/Coastal Discussion Paper.

FEPS 5. Baird & Associates. Beach Access Performance Indicator Summary.

FEPS 6. Memorandum from Zuzek, P.J., and R. Roblin (Baird & Associates) to IJC Study Participants re: Integration of Beach User Economics into the Shared Vision Model. April 7, 2004.
FEPS 7. Baird & Associates. March 2004. Erosion Performance Indicator: Methodology and Shared Vision Model Application. Prepared for I.J.C. Plan Formulation and Evaluation Group.
FEPS 8. Baird & Associates. February 2005. Flooding Performance Indicator: Methodology and Shared Vision Model Application. Prepared for IJC Plan Formulation and Evaluation Group. Oakville, Ontario: Baird & Associates.
FEPS 9. Baird & Associates. February 2004. Shore Protection Performance Indicator: Methodology and Shared Vision Model Application. Oakville, Ontario: Baird & Associates.
FEPS 10. Barrier Beaches and Dunes PI Summary.
FEPS 11. Contextual Narrative for "Sediment Budget PI" Lake Ontario and the Upper St. Lawrence River.
FEPS 12. Contextual Narrative for Erosion, Flooding, and Existing Shore Protection PIs Lake Ontario and the Upper St. Lawrence River.
FEPS 13. Contextual Narrative for "Beach Access" and "Barrier Beaches & Dunes PIs" Lake Ontario and the Upper St. Lawrence River.
FEPS 14. Baird & Associates. Erosion Performance Indicator Summary.
FEPS 15. Baird & Associates. Flooding Performance Indicator Summary.
FEPS 16. Sediment Budget PI Summary.
FEPS 17. Baird & Associates. Existing Shoreline Protection Performance Indicator Summary.
FEPS 18. Minutes/Observations from FEPS Independent Technical Review 12/13/2002 and 7/26/2002 notes.
Hartford, D. N. D., and G. B. Baecher. 2004. Risk and Uncertainty in Dam Safety. London: Thomas Telford.
IJC (International Joint Commission) International Red River Basin Task Force. 2000. The Next Flood: Getting Prepared. Ottawa and Washington: Final Report to the International Joint Commission
Morgan, M. G., and M. Henrion. 1990. Uncertainty: A Guide to Dealing with Risk and Uncertainty in Quantitative Risk and Policy Analysis. New York: Cambridge University Press.
Nairn, R.B., and H. N. Southgate. 1993. Deterministic profile modeling of nearshore processes: Part 2, Sediment transport and beach profile development. Coastal Engineering 19:57-96.
Nairn, R. B., P. Zuzek, A. Morang, and L. E. Parson. 1997. Effectiveness of Beach Nourishment on Cohesive Shores, St. Joseph, Lake Michigan. USACE Technical Report CHL-97-15.
NIST (National Institute of Standards and Technology). 2005. Engineering Statistics Handbook. Available on-line at *http://www.itl.nist.gov/div898/handbook/*. Accessed October 2005.
NRC (National Research Council). 2000. Risk Analysis and Uncertainty in Flood Damage Reduction Studies. Washington, DC: National Academy Press.

NRC. 2001. Under the Weather: Climate, Ecosystems, and Infectious Disease. Washington, DC: National Academy Press.

Parson, L. E., A. Morang, and R.B. Nairn. 1996. Geologic Effects on Behavior of Beach Fill and Shoreline Stability for Southeast Lake Michigan. USACE Technical Report CERC-96-10. Vicksburg, MS: U.S. Army Corps of Engineers Waterways Experiment Station.

Pilkey, O.H., and J. A.G. Cooper. 2004. Society and sea level rise. Science 303:1781.

Rescher, N. 1998. Predicting the Future: An Introduction to the Theory of Forecasting. New York: State University of New York Press.

Resio, D., and W. Perrie. 1989. Implications of an F-4 equilibrium range for wind-generated waves. Journal of Phys. Oc. 19(2):193.

Scott, D, D. Schwab, P. Zuzek, and C. Padala. 2004. Hindcasting wave conditions on the North American Great Lakes. Available on-line at *http://www.ocean-weather.com/8thwave/Papers/D4.pdf*. Accessed October 2005.

Simonovic, S.P., and R.W. Carson. 2003. Flooding in the Red River Basin–Lessons from post flood activities. Natural Hazards 28:345-365.

Southgate, H. N., and R. B. Narin. 1993. Deterministic profile modeling of near-shore processes: Part 1, waves and currents. Coastal Engineering 19:27-56.

USACE (U.S. Army Corps of Engineers). 2003. Coastal Engineering Manual–Chapter 6, Reliability based design of coastal structures EM1110-2-1100 (Part VI).

Voortman, H. G., P. van Gelder, and J. K. Vrijling. 2002. Risk-based design of large-scale flood defence systems. 28th International Conference on Coastal Engineering. Pp. 2373-2385 In Solving Coastal Conundrums. July 7, 2002-July 12, 2002. Cardiff.

Zuzek, P. J. 2005. Presentation to the NRC Committee to Review the Lake Ontario-St. Lawrence River Studies. June 14 and June 15, 2005. Niagara-on-the-Lake, Ontario, Canada.

Chapter 4
Environmental Sciences in the LOSLR Study: Wetlands, Species at Risk, and the Integrated Ecological Response Model

The environmental sciences work presented for NRC review includes three subjects: wetlands, species at risk (SAR), and the integrated ecological response model (IERM). Because of the strong linkages and similar issues and strengths among them, these three are evaluated in this chapter as a collective body of environmental science work in the LOSLR Study. The first section of the chapter introduces the environmental science studies and models of the LOSLR Study and notes their accomplishments. The second section evaluates the environmental work in terms of how well they reflect reasonable scientific methods, assumptions, and supported findings. The next section continues the evaluation in terms of how key information is displayed and integrated to understand tradeoffs among candidate regulation plan options. The last evaluative section notes the degree to which the environmental sciences, collectively, are appropriate or sufficient to be used to evaluate regulation plans and determine impacts of water level changes. The chapter concludes with cross-cutting comments and specific summaries for the wetlands, SAR, and IERM work, respectively.

BACKGROUND AND FOCUS OF ENVIRONMENTAL SCIENCES WORK IN THE LOSLR STUDY

The Moses Saunders dam and the current operating plan (Plan 1958DD) have modified the natural dynamics of discharge and water-level fluctuations in Lake Ontario, the St. Lawrence River, and the associated lacustrine and riverine wetlands (IERM 1). These modifications are viewed as contributing to undesirable ecological impacts of the Lake Ontario–St. Lawrence ecosystem (IERM 1, page 3). In response to these and other environmental concerns and as part of an updated plan for water-level and flow regulation, a guiding principle of the IJC Study Board was that "Criteria and Regulation Plans will contribute to the ecological integrity of the Lake Ontario-St. Lawrence River ecosystem" (IERM 1, page 3). The level of funding from Canada and the United States underscores the importance that the IJC places on environmental issues in the LOSLR system. The Study Board's efforts to incorporate environmental concerns into a regulatory process that most often and acutely focuses on economic and engineering considerations represent one of the notable aspects of the LOSLR Study.

Members of the Environmental Technical Working Group developed and undertook a field research program and literature review to predict the response of selected organisms or groups of organisms (e.g. wetland plants, fish, herptiles, birds and SAR) to water levels and river flows. The wetland, faunal, and SAR field studies led to the development of algorithms that describe relationships among water level, river flow, organisms, and communities of concern along the Lake Ontario shoreline and up- and down-stream of the Moses Saunders Dam. Organisms in these field studies include an array of birds, fish, herptiles, wetland plant communities and SAR.

The Environmental Technical Working Group also oversaw the development of the Integrated Ecological Response Model (IERM) for ecological assessment of different management plans for the lake and river. Thirty-two Performance Indicators (PIs) were identified to represent the LOSLR system's ecological response to water levels and flows. The IERM was developed by Limno-Tech, Inc. and Environment Canada to work within the SVM to facilitate comparisons between alternate plans on environmental issues relative to Plan 1958DD.

Documents Presented for Review

The IJC presented the committee with several documents that represent the state of the environmental sciences work. What follows is a list of the documents presented for review; the complete list of documents, including background documents and those distributed at committee meetings, are listed in Appendix B.

Wetlands Documents:

- ***W 1*** – Wilcox, D. A., K. P. Kowalski, and M. L. Carlson. 2003. Year 3 Report of Relationship between Lake Ontario Water Levels and the Wetlands. Ann Arbor, MI: Great Lakes Science Center, U.S. Geological Survey.
- ***W 2*** – Evaluation of Water Level Regulation Influences on Lake Ontario and Upper St. Lawrence River Coastal Wetland Plant Communities.

IERM Documents:

- ***IERM 1*** – Descriptive Documentation; Limno-Tech, Inc. April 2005. Draft Development of an Integrated Ecological response Model (IERM) for the Lake Ontario-St Lawrence River Study.
- ***IERM 2*** – IERM Model (version 4.1.1)

SAR Documents:

- ***SAR 1*** – Water Fluctuation Impacts on Species At Risk Cornwall to Pointe due Lac [French]
- ***SAR 2*** – Lake Ontario Species at Risk Supplement (Least Bittern and Black Tern Reproductive Index Performance Indicators)
- ***SAR 3A*** – Impact of Water Level Regulation on Nearshore Habitat Availability and SAR
- ***SAR 3B*** – (Supplement) Impact on dunes and SAR
- ***SAR 4*** – Species at Risk Fish Supplement (Year 4 Modeling Group Report)

- ***SAR 5*** – Lower St. Lawrence Species at Risk Final Report (2005) [French]

ACCOMPLISHMENTS OF THE ENVIRONMENTAL SCIENCES IN THE LOSLR STUDY

One strength of the LOSLR study is the plant community composition analysis at shoreline elevations associated with distinct water level histories. These inventories were repeated in 32 sites around Lake Ontario. The field research appropriately partitioned the sampling site selection across four common hydrogeomorphic wetland types of the Great Lakes: open embayment, protected embayment, barrier beach, and drowned river mouth. Included in the wetland plant studies is a historic imagery analysis of the vegetation communities that have resulted from significant flooding or dewatering events over the last 40 years. Previous Great Lakes research demonstrates that coastal wetlands respond to flooding and the water level of a lake (see summary in Keough et al., 1999).

The SAR studies proved to be a valuable contribution to the state of knowledge of the LOSLR system. They collate an enormous amount of natural history information on species in the study area that often have legal protection status (e.g., species reports in Appendix B of SAR 3A). This information is not readily accessible to the scientific world or the general public.

The development of Target Plots was handled well. The presentation of environmental data using the "Target Plot" was effective in its easy and informative grouping of the Performance Indicator ratios by type (fish, birds, etc.). Target plots allowed a quick but informative method to visualize the impact of a water level plan.

Finally, the speed with which the models were created was impressive. In five short years, a series of wetland models were developed based on empirical data of one of the largest and most complicated aquatic ecosystems in the world. Unlike the hydraulic, flow, and erosion models that have had decades of research and development, application and refinement, ecological models, like the IERM and its sub-models, are in a relatively early stage of refinement and testing. This feat alone will likely provide impetus to ecological science, especially the fields of ecological modeling and restoration ecology, if the science behind the IERM is published in peer-reviewed journals.

EVALUATION OF SCIENTIFIC METHODS, ASSUMPTIONS, AND SUPPORTED FINDINGS OF THE LOSLR ENVIRONMENTAL SCIENCES

The first charge (Charge A) in the statement of task asks the degree to which the studies and models reflect reasonable scientific methods, assumptions, and supported findings. This charge is addressed through three evaluation criteria: empirical foundations, quality assurance, and treatment of risk and uncertainty.

Criterion 1: Empirical Foundations

Empirical foundations of the environmental sciences form the basis for faunal sub-models, environmental Performance Indicators, and ultimately environmental tradeoffs in the SVM. They are discussed in terms of wetland sampling, SAR sampling, missing empirical data, performance indicators, and the ecosystem v. single-species approach to environmental science.

Wetlands Sampling

The wetland vegetation analysis identified four distinct vegetation communities associated with suites of flood/dewater histories within 32 wetlands distributed primarily around the eastern half of Lake Ontario in Canada and the United States: the marsh meadow (not flooded 5-30 years), marsh meadow/emergent mix (not flooded < 5 years or dewatered < 4 years), emergent marsh (not dewatered 4-39 years), and submergent/floating aquatic bed (not dewatered 40 years or more). The Environmental Technical Work Group chose to focus on the marsh meadow as the sensitive indicator. Although undocumented, there are several possible reasons for this decision: the marsh meadow habitat is high in species richness; it is a critical habitat for some wildlife; and the marsh meadow habitat is a typical vegetation type seen in all of the Great Lakes.

Because detailed bathymetry/topography data were not available for the majority of the inventory (IERM 3, pg. 12), physical data collected for the 32 sites were used to derive "typical" wetland geometry for each type. The Environmental Technical Working Group supported this decision based on perceived consistency in the bathymetric configuration within wetland geomorphic types. The total area of each plant category for each geomorphic type was calculated by "simply multiplying the percent cover by the total area of each geomorphic type in the system (W 1)." Validity of this approach implicitly requires that sampled wetlands be (1) selected randomly and (2) demonstrated to be representative of those in the inventory for Lake Ontario and the Upper St. Lawrence River. The key wetland and IERM documents do not indicate how the wetlands were selected for sampling. The 32 "typical" wetlands do not appear to have been chosen randomly nor were they sampled to ensure representativeness of available wetlands by type. Therefore, the validity of applying study results to the entire wetland inventory remains in question.

Table 4-1 was prepared using the W1 Appendix A to derive a summary of wetland size and numbers for Lake Ontario and the St. Lawrence River and to examine the degree to which the study wetlands may be representative of the population of wetlands. Figure 4-1 illustrates the distribution of wetland types by area and number along the Canadian and US coast of Lake Ontario and along the St. Lawrence River.

There are very few open embayments in Lake Ontario, and most of these occur in the U.S. and along the river (Figure 4-1). The average size of wetlands in the U.S. ranges from 6 ha for open embayments to 35 ha for barrier-beach wetlands. By comparison, the average size of Canadian wetlands is much larger, 33 ha for barrier-beach to 66 ha for drowned river mouth marshes. The average size of riverine wetlands tends to be intermediate, from 21 to 39 ha (Table 4-1). Compared with what is available, the average size of wetlands sampled in this project was between 19 and 46 ha, generally much larger than the average wetland in the United States, except for the barrier-beach type (Table 4-1). Unfortunately, the degree to which these untested assumptions of representativeness contributed to error propagation is unknown. **A thorough analysis of the wetland study should document how well the sampled sites represent wetlands across Lake Ontario and the St. Lawrence River**

By consistently sampling only four of each wetland type in each country, the study may have inadvertently biased results in favor of open and protected embayments (see Table 4-1). Simultaneously, this approach may have under-sampled drowned river mouth marshes, especially in the heavily urbanized west end and north shore of Lake Ontario. A rough estimate using the inventory (Appendix A of W2) showed 37 wetlands that could be described as urbanized, and these accounted for 970 ha. In addition, wetland size should be considered in any sampling scheme as wetland area may be an important factor in vulnerability to perturbation and may be important to a site's function as wildlife habitat.

Wetlands sampling was limited to a minimum depth of 74.25 m above sea level (asl) (IERM 3, p. 12). However, the typical profiles developed for the four geomorphic types included depths from 73.00 to 75.75 m asl (IERM 1 p. 21). From the existing documentation, it is difficult to ascertain how the Environmental Technical Working Group assigned plant information to depths between 73.00 and 74.25 m. This sampling regime appears to have under-sampled the submersed aquatic vegetation (SAV), since sampling only included depths <1.0 m. By not sampling in deeper waters, the study excluded many of the submergent species in the high-quality wetlands, and this may explain why there are fewer than 10 common SAV species in Tables 4, 6, 8 and 10 in W2. Had the study expanded its sampling effort to survey more of the deeper aquatic habitat, a more representative suite of species of submergent species would likely have been reported. **The current sampling methods likely underestimated the diversity of submersed vegetation and available fish habitat.**

Many of the habitat models used to develop the SAR Performance Indicators rely on predictions of wetland vegetation, including submergent vegetation. All of these Performance Indicator models will be vulnerable to the same criticisms that have been made of the wetland vegetation models and vegetation sampling methods, such as sampled wetlands are not statistically representative of wetlands in the LOSLR; submergent vegetation appears to be undersampled; and there are limited linkages between bathymetric changes and vegetation models. **Error from wetland sampling deficiencies may be propagated through the SAR material, through the IERM, and into the SVM. This is a serious limitation to the credibility of these SAR Performance Indicators.**

TABLE 4-1 LOSLR Wetland Study Sites in U.S. and Canada in Comparison to Total

Category	Barrier Beach	Drowned River Mouth	Open Embayment	Protected Embayment
All U.S. wetlands (ha)	4,046.25	1,961.01	386.2	937.97
Total # US wetlands	117	99	67	29
All Canadian wetlands (ha)	2,485.56	5,908.06	1,235.89	2,915.49
Total # Canadian wetlands	76	90	20	55
All SL River wetlands (ha)	469.98	1,288.17	1,714.41	2,498.20
Total # SL River wetlands	26	44	73	183
All wetlands in LO (ha)	6,531.81	7,869.07	1,622.09	3,853.46
Total # wetlands in LO	193	189	87	84
All LO + SL river wetlands (ha)	7,001.79	9,157.24	3,336.5	6,351.66
Total # wetlands (LO + SL River)	219	233	160	267
Average size of wetlands in US (ha)	34.58	19.81	5.76	32.34
Average size of wetlands in Canada	32.71	65.65	61.79	53.00
Average size of riverine wetlands (ha)	31.97	39.30	20.85	23.79
Average size of wetlands sampled in LO	33.84	41.63	18.65	45.87
Area of wetlands sampled in LO (% of total)	837 (12.8%)	1,443 (18.3%)	882 (54.4%)	949 (24.6%)
No. of wetlands sampled in LO (% of total)	8 (4.1%)	8 (4.2%)	8 (9.2%)	8 (9.5%)
No. of wetlands sampled in US (% of total)	4 (3.4%)	4 (4.0%)	4 (6.0%)	4 (13.8%)
No. of wetlands sampled in Canada (% of total)	4 (5.3%)	4 (4.4%)	4 (20.0%)	4 (7.3%)

Note: Regrouping of original data presented in Appendix A of **W2** to evaluate the degree to which the 32 study sites were representative of available wetlands in both countries and in the St. Lawrence River. Numbers in bracket are percent (%) of total wetlands in all of Lake Ontario, or only in the Canadian or U.S. portions of Lake Ontario, as appropriate. Statistics on sampled wetlands were taken directly from **IERM 3** (Table 2, p. 12).

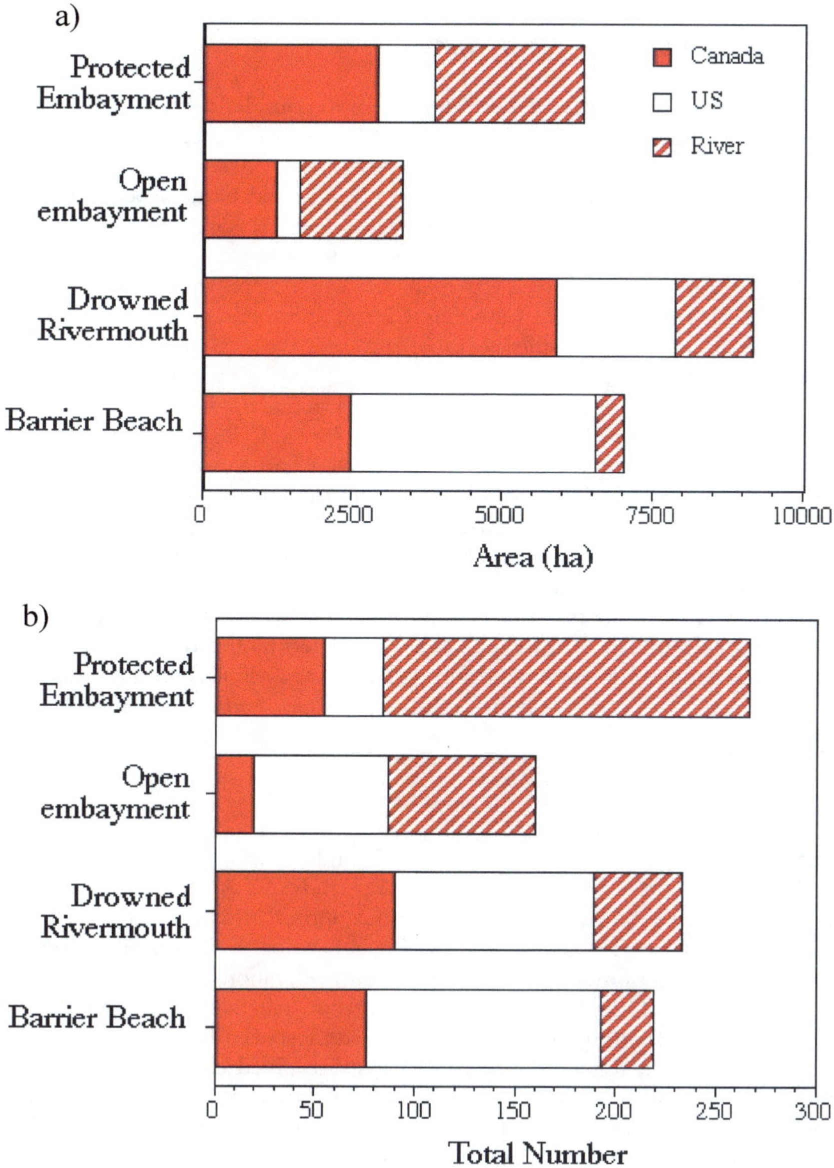

FIGURE 4-1 Distribution of wetlands by geomorphic types in Lake Ontario and the St. Lawrence, with respect to a) total area and b) total number.

External Environmental Stressors

The models and Performance Indicators seem to consider wetlands as discrete elements, when in fact, wetlands reflect attributes of their surrounding aquatic and upland systems. Surrounding uplands areas can be a major determinant of wetland structure and function through impacts to the physical and biotic aspects of wetland ecology. Without taking these external forces and stressors into account, the IERM cannot be used to predict the combined effects of water-levels and other stressors such as deterioration in water quality, adverse impacts of exotic species (Lougheed et al., 2004; Wei and Chow-Fraser 2006) and urban encroachment, that are known to affect the re-establishment of different types of aquatic vegetation (Chow-Fraser et al., 1998; Wei and Chow-Fraser, 2005). For example, there are heavily urbanized wetlands in western and northern Lake Ontario (see Cootes Paradise Marsh in Chow-Fraser 2005; Frenchman's Bay in Eyles et al., 2003), where currently there are no wet meadow communities due to urban encroachment. The sampling program could have been additionally stratified to account for the effects of altered land uses (i.e. urban and agricultural development) on plant distribution, so that results could have been incorporated into an adaptive management model.

In all, the wetlands sampling effort was broad, but should have been carried out to reflect availability of the four geomorphic types rather than having all four types being equally represented in the study set. The number of wetlands to be sampled should have been determined through a power analysis, and then the sites should have been chosen randomly from the Lake Ontario wetland inventory.

SAR Sampling

The field data collected during the LOSLR studies are very limited and were obtained only over a few field seasons (see SAR 1 and SAR 3A for lists of species at risk considered for the study). The limitations appear in several aspects of the SAR work. For example, many SAR exhibit substantial inter-annual variation, which would not be captured by field studies of short duration (and, in some cases, single visits to a site). Other SAR were excluded from analysis because they do not occur in the study area currently. Intuitively, excluding species from analysis would seem correct; however, this may prove to be a shortsighted choice since a stated purpose in developing a new water regulation plan is to reduce environmental degradation due to water level management and, presumably, restore some missing species. Certainly, it is more difficult to predict responses to water level management for species which have not been observed in the study area, but this may be no greater a limitation than reliance on only a handful of observations, which is already the case for some of the SAR for which Performance Indicators have been developed.

Performance Indicators

Performance Indicators employed in the LOSLR environmental science work are often indirect measures of integrity of a community such as wetland area. In most cases, it is not clear why these metrics were chosen when more conventional, ecological assays, such as diversity, biomass, or productivity remain viable metrics. For example, the various plant Performance Indicators are based on the areal depth available for growth (available suitable habitat) for a community at a given lake level. Implicit in this Performance Indicator is the assumption that lake level area is directly related to species abundance, productivity or species diversity. Defense of this approach is needed in the environmental documentation because a large body of research shows that a number of different physical and biotic factors in addition to hydrology affect species distribution and growth (Hupp, 2000; Naiman et al., 1993; Pollock et al., 1998).

Two different approaches were used to develop performance indicators (Table 4-2) for the wetland sub-models, one for the Lake Ontario-Upper St. Lawrence, and another for the Lower St. Lawrence (IERM 1, p.12). The Lake Ontario-Upper St. Lawrence approach (IERM 1 p.12) relied on information from 32 "typical" wetland sites that represented equally the four geomorphic types (barrier beach, drowned river mouth, open embayment and protected embayment). The wetland Performance Indicator reflects the average total area of each vegetation community (specifically meadow marsh; IERM 1, p. 24) that occurred in the 32 sample sites. These sites were extrapolated to an inventory of wetlands within Lake Ontario-Upper St. Lawrence that were classified according to geomorphic type. By comparison, the Performance Indicator for the Lower St. Lawrence was based on a comprehensive digital elevation model that covered the Lower St. Lawrence channel and the entire floodplain for Lake St. Louis and the reach from Montreal Harbor to Lake St. Pierre. The model framework was linked to a flexible database system, which allowed calculation of relevant physical and biological information including vegetative cover of treed swamp, shrubby swamp, prairie meadow, shallow marsh, deep marsh, open water, and invasive prairie meadow, total wetland area, as well as surface area of several ecologically important submergent aquatic plant species (IERM 1, p. 18). Node results computed by the 2-D model were stored in the database system and were aggregated to provide summaries of model output for key locations in the LSL system.

The dune-barrier beaches report nicely describes the dune-barrier beaches; it even documents the report's shortcomings. It was not clear why there was no Performance Indicator developed to represent this habitat or its characteristic species, many of which are SAR. Failure to include barrier beaches and dunes misses an opportunity to consider species and habitats associated with them, particularly along the southern coast of Lake Ontario, where loss of barrier beaches and dunes will also jeopardize wetlands that are presently protected by them.

TABLE 4-2 Key Ecological Performance Indicators

SVM ID	Performance Indicator Group	Description (Region)	Units
E1	Vegetation	Wetland Meadow Marsh Community–total surface area and supply based (Lake Ontario)	ha
E2	Fish	Low Veg 18C – spawning habitat supply (Lake Ontario)	ha-days
E3	Fish	High Veg 24C – spawning habitat supply (Lake Ontario)	ha-days
E4	Fish	Low veg 24C – spawning habitat supply (Lake Ontario)	ha-days
E5	Fish	Northern Pike – YOY recruitment index (Lake Ontario)	index
E6	Fish	Largemouth Bass – YOY recruitment (Lake Ontario)	index
E7	Birds	Virginia Rail (RALI) median reproductive index (Lake Ontario)	index
E8	SAR	Least Bitten (IXEX) – median reproductive index (Lake Ontario)	index
E9	SAR	Black Tern (CHNI) – median reproductive index (Lake Ontario)	index
E10	SAR	Yellow Rail (CONO) – preferred breeding habitat coverage (Lake Ontario)	ha
E11	SAR	King Rail (RAEL) – preferred breeding habitat coverage (Lake Ontario)	ha
E12	Fish	Low Veg 18C – spawning habitat supply (Upper St. Lawrence)	ha-days
E13	Fish	High Veg 24C – spawning habitat supply (Upper St. Lawrence)	ha-days
E14	Fish	Low Veg 24C – spawning habitat supply (Upper St. Lawrence)	ha-days
E15	Fish	Northern Pike – YOY recruitment index (Upper St. Lawrence)	index
E16	Fish	Largemouth Bass – YOY recruitment (Upper St. Lawrence)	index
E17	Fish	Northern Pike – YOY net productivity (Upper St. Lawrence)	grams/ha
E18	Birds	Virginia Rail (RALI) – median reproductive index (Lake St. Lawrence)	index
E19	Mammals	Muskrat (ONZI) – house density in drowned river mouth wetlands (Thousand Islands area)	number/ha
E20	Fish	Golden Shiner (NOCR) – suitable feeding habitat surface area (Lake St. Louis to Trois-Rivières)	ha
E21	Fish	Wetlands fish – abundance index (Lower St. Lawrence)	index
E22	Fish	Northern Pike (ESLU) – suitable reproductive habitat surface area (Lake St. Louis to Trois-Rivières)	ha
E23	Birds	Migratory wildfowl–floodplain habitat surface area (Lake St. Louis to Trois-Rivières)	ha
E24	Birds	Virginia Rail (RALI) – reproductive index (Lake St. Louis to Trois-Rivières)	index

continues

TABLE 4-2 Continued

SVM ID	Performance Indicator Group	Description (Region)	Units
E25	Birds	Migratory wildfowl – productivity (Lake St. Louis to Trois-Rivières)	number of juveniles
E26	Birds	Black Tern (CHNI) – reproductive index (Lake St. Louis to Trois-Rivières)	index
E27	Herptiles	Frog sp. – reproductive habitat surface area (Lake St. Louis to Trois-Rivières)	ha
E28	Mammals	Muskrat (ONZI) – surviving houses (Lake St. Louis to Trois-Rivières)	number of houses
E29	SAR	Least Bittern (IXEX) – reproductive index (Lake St. Louis to Trois-Rivières)	Index
E30	SAR	Easter San Darter (AMPE) – reproductive habitat surface area (Lake St. Louis to Trois-Rivières)	ha
E31	SAR	Spiny Softshell Turtle (APSP) – reproductive habitat surface area (Lake St. Louis to Trois-Rivières)	ha
E32	SAR	Bridle Shiner (NOBI) – reproductive habitat surface area (Lake St. Louis to Trois-Rivières)	ha

Only limited information on Performance Indicator models is presented in the English summaries of SAR 1 and SAR 5. In one of three cases where the same SAR is included in both the lower river (SAR 1 and 5) and upper river and lake documents (SAR 2) (Least Bittern), it appears that the structure of the models for both regions may be similar. It is not clear whether separate data or separate expert judgments were used to parameterize the models for different regions. In the other two cases (Spiny Softshell Turtle in SAR 5 and SAR 3A, and Bridle Shiner in SAR 5 and SAR 4), the Performance Indicator models appear to be different for the two regions, but it is unclear whether different structures are based on differences in data availability or differences in habitat responses in the two regions. Furthermore, Performance Indicator ratios are calculated for rate processes, biomass, area, etc., using different units of measure. Nowhere in the documentation are these different units reconciled, resulting in different Performance Indicators being based on different units. **A discussion is needed about how the calculation of Performance Indicator ratios using different base units (biomass, rate process, etc.) may affect error statement and certainty.**

The Environmental Technical Working Group collapsed several hundred Performance Indicators in the IERM to 32 (Table 4-2). The collapse to 32 Performance Indicators is viewed as reasonable, but the rationale behind the number 32 needs to be explained. In the illustrative case of the SAR Performance Indicators, it is sensible to reduce the original list of Performance Indicators to concentrate on those most likely to be affected by water level regulation, and then to narrow that list further to retain as indicators SAR that can "represent" a suite of other species expected to respond similarly. However, the criteria used to reduce the lists at each stage need to be stated more explicitly. The reductions also need to be justified in terms of what is known about the natural history and population dynamics of the SAR. These issues are discussed in Appendices C and D of SAR 3A for Lake Ontario and Upper St. Lawrence SAR, but are lacking from most of the other SAR documents. **Because of issues of validation, error propagation and sensitivity, the use of all 32 Performance Indicators (32) would provide a better interpretation of environmental effects for a plan, and the reasons for deciding against the larger number of Performance Indicators should be better explained.**

The environmental Performance Indicators were designed to allow relative rankings of plan alternatives (ordinal measurement scale) rather than to predict absolute levels of SAR population responses to water level regulation (ratio measurement scale). This is a sensible goal, particularly given the limited data and limited understanding of the needs of many SAR. However, when they are subsequently integrated with the other metrics used to evaluate alternatives, **environmental Performance Indicators must also provide information on how much better or how much worse one plan is compared to another (interval measurement scale); otherwise, it will not be possible to evaluate tradeoffs among environmental Performance Indicators and between environmental Performance Indicators and those that are expressed in monetary units.**

Ecosystem vs. Single-Species Approach

In general, the wetlands work and the IERM use a single-species approach rather than an ecosystem approach. A single-species approach focuses on the reaction of an organism to lake level or river flow. An ecosystem approach considers processes that transcend individual species and indicate integrity of the ecosystem as a whole, such as supra-specific measures like total site biomass and net primary production. An ecosystem approach can also account for the presence and abundance of organisms, diversity and composition of plant and animal communities, and their reactions to other stressors in the LOSLR environment, such as contaminants, inter-specific competition, invasive species, nutrients and sediment supplies from the watershed, climate change, and other fisheries stocking and harvesting practices. Omission of the effects of these other stressors in the LOSLR study may compromise the interpretation provided in the Shared Vision Model. The current design of the IERM (and the SVM) makes it very difficult to separate the effects of these and other stressors from effects due to water regulation alone.

Sometimes, ecosystem-oriented approaches fail to account fully for species with special requirements not captured in the ecosystem-level model. For situations where field data on individual species requirements are limited, as is the case for many of the SAR in this study area, protecting and monitoring essential ecosystem features (such as dune processes) may be an effective and efficient strategy for protecting SAR (Franklin, 1993). Monitoring ecosystem features, supplemented by selected monitoring of individual SAR status, would provide the feedback necessary to assess the merits of whichever water regulation plan is implemented.

There are two Performance Indicators that embrace an ecosystem approach.

The wetland marsh meadow Performance Indicator is one, as it is a community metric with clear relationships to environmental conditions. Another example of an ecosystem approach in the environmental documents is the report on dunes and SAR for the Eastern Lake Ontario Barrier Complex (SAR 3B). This document supports an ecosystem-oriented, rather than single-species oriented approach to evaluation of impacts on SAR. **An ecosystem-approach is recommended with supplemental monitoring of individual species.**

Criterion 2: Quality Assurance and Quality Control

Quality Assurance/Quality Control (QA/QC) in the environmental sciences provides ways to validate study and model results and a means to ensure that results are accurate and reliable. QA/QC does not appear to have been instituted for the LOSLR Study. In general, information on validation, calibration and verification presented on various algorithms in models is weak or non-existent for wetlands, SAR, and the IERM. Initial formal direction from the Study Board on QA/QC could have reduced concerns about the lack of validation and significance-testing that pervades the work in this component of the LOSLR study.

Validation

IJC materials did not describe any independent validation of the IERM in Lake Ontario, the St. Lawrence River, or elsewhere in the Great Lakes. In fact, the statements are repeatedly made in reviewed documents that "No calibration data are available," "No specific validation data sets are available" (IERM 3). During the question/answer sessions with the Environmental Technical Working Group representatives at the first committee meeting, no further information was provided about calibration or validation exercises. This lack of validation makes it impossible to comment on the accuracy of predictions.

An assumption is built into the IERM model that cattails will be reduced with draw-downs over time. This hypothesis needs to be experimentally verified to complete the wetland investigations. This verification is keenly needed because there is published evidence to the opposite effect: cattail distribution has been shown to increase with low water levels in Cootes Paradise Marsh, a large urban marsh at the western end of Lake Ontario (Chow-Fraser 2005; Wei and Chow-Fraser 2005; 2006). Literature also shows that *Typha* species have broad environmental tolerance and are excellent competitors. Recent expansion of invasive species, such as *Phragmites australis,* into the sedge meadow under similar conditions have also been noted (W1). Both of these situations would result in an expanded emergent rather than sedge-meadow zone. The marsh meadow hypothesis needs additional testing to be validated. **The previous IJC Water Levels Reference Study (early 1990s) (Wilcox et al., 1993) may provide valuable data that could be used to validate the current Lake Ontario wetland results, and such test-**

ing is needed to demonstrate the quality of the analysis and the certainty of forecasts.

Aerial photo analyses were used as part of the wetland empirical exercises. Aerial photographs indicate that emergent marsh and especially cattail (*Typha angustifolia*) are favored by flooded conditions and by relatively stable inter-annual patterns of lake level and climate in Lake Ontario. The study would benefit from a quantitative analysis of the historic series of aerial photographs.

For most SAR modeled, there are no data available for quantitative model validation and no qualitative validations are offered (other than the statement "accurate for relative comparisons," without further elaborations). In the few cases where quantitative validation was attempted, the results often are not presented clearly or are not convincing. For example, the prediction rate for Least Bittern nest habitat is stated to be 75.6 percent, but it is not clear what "prediction rate" means. For Yellow Rail, two nests were available for validation. One nest 215 m away from predicted nesting habitat is described as "very near" and the other was 375 m away. This "near" value is not convincing.

The quality of the wetland results would be strengthened if the authors had provided more information on the precision of elevation measurements, water-level measurements (including seiche effects), and the accuracy and precision of vegetation measurements and identifications. **Although the vegetation zones are sufficiently distinct and the relationship to flooding sufficiently strong that some variation would not likely have changed the results, an explicit explanation of sources of variation is needed to demonstrate confidence in the results.**

Inconsistent Terminology

Variable use of terminology presented confusion in some of the environmental science documents. Prediction of wetland area as a function of water level was verbally reported as excellent; uncertainty for the algorithms for fish, birds and SAR was verbally reported as high. In no place in the documents are "excellent" or "high" given quantitative values. A substantive set of concerns was raised by the use of the term "precautionary principle." The phrase is used in a non-standard way in the confidence ratings in SAR 5, with no apparent relation to the usual meaning in the literature (Lauck et al., 1998). The same phrase is used in SAR 2 to give "conservative" Performance Indicators, but the meaning of this statement is not clear. Based on verbal descriptions of "conservative," the approach taken may actually be the opposite of the understanding of the precautionary principle as applied to environmental decision making in the literature. Other terms present confusion in the environmental documents, as well. SAR Performance Indicator models refer to "rapid" rises in water levels, but "rapid" is not defined. Likewise, "stranded" is used in the wetlands documentation, but needs further clarification, in terms of how stranding occurs under different water level scenarios and how these impact species of interest, such as the northern pike.

Environmental Technical Working Group scientists provided ratings of confidence in the IERM and some sub-models. Confidence ratings are important, standard practice. Confidence ratings are particularly useful in cases that expert opinion helped develop habitat models, as in this case. Unfortunately, in this case, the ratings provided appear to be largely "boilerplate," with the same phrases repeated for each species and with no evidence offered for the ratings being "accurate for relative comparison." For example, the Yellow Rail Performance Indicator is rated as "allowing for relative comparison" and, in the same paragraph, as "should not be used as a key indicator" due to limitations in information or partial failure of validation. Without explanation, these judgments appear contradictory. In the appendices of SAR 3A, similar discussions are labeled "Risk and Uncertainty Assessments," rather than "Confidence."

SAR 2 and SAR 5 appear to come to the opposite conclusion based on similar types of models and data. Such contradictory assessments illustrate the need for well-defined and consistent standards for judging reliability, especially when those judgments are qualitative. It is helpful to comment on the sensitivity of the SAR Performance Indicators to water level regulation, as is done in the appendices to SAR 3A for at least some SAR. **Terminology should be consistent with standard usage, defined and clarified in the environmental sciences documentation.**

Criterion 3: Treatment of Error and Uncertainty

Like quality control issues, a formal method to test for error and uncertainty was not documented for field investigations or simulations. In general, information on error propagation, risk, and uncertainty is weak or non-existent for SARs, wetlands and the IERM, and IJC documents note that "…cumulative uncertainties have not been estimated" (IERM 3). Deficiencies in the treatment of error and uncertainty were identified with respect to methods used in wetland sampling, SAR analyses, and creation and selection of Performance Indicators (see Criterion 1: Empirical Foundations).

IJC documents describe a 10 percent change rule to convey significant variability associated with environmental Performance Indicators. No written documentation exists on how the 10 percent rule of significance was achieved, although verbal comments in the committee meeting open session suggested that it was selected without scientific or practical justification. According to the oral comments, ecological index and Performance Indicator ratios are based on the "board's professional judgment," and not derived from traditional statistical methods. An "administrative decision" was made to accept a 10 percent change as a significant difference for the environmental indicators. In other words, LOSLR staff is using this 10 percent rule to assign value to an effect size that is best described as a non-statistical choice about what difference is big enough to influence a particular decision. This is an area of major uncertainty for the decision making process in the SVM via the IERM. **A better explanation about error and uncertainty is**

needed to convey confidence in the environmental sciences results for LOSLR decision making.

Because decisions about water level regulation will be based, in part, upon the environmental sciences information, the IJC needs a high level of confidence in the conclusions derived from the myriad of models produced. The environmental sciences work presents an illustrative example of how error could be introduced and propagated throughout the LOSLR study structure. Since the faunal sub-models for fish, wetland birds, muskrat, and herptiles were completely or partially dependent on habitat information derived from the wetland plant sub-model, the wetland Performance Indicator strongly impacts all those models and assumptions based upon it. Flaws in empirical foundations—wetlands representativeness, missing or limited empirical data—will influence the IERM and the predicted outcomes of the SVM. Without a clear discussion or description of estimates of error or uncertainty, the degree to which the SVM reflects these flaws cannot be determined. **The overarching concern of error and uncertainty in the environmental sciences is how error is propagated from the wetlands and SAR sampling methodologies into the Performance Indicator development and through use in the IERM and SVM.**

INTEGRATION AND DISPLAY OF KEY INFORMATION

The second charge (Charge B) in the statement of task looks to understand how well the studies and models integrate and display key information needed for a comprehensive evaluation and understanding of the tradeoffs for selecting among the candidate RPOs. Four evaluation criteria are used to address this charge: linkages among related studies and models; temporal scaling resolution; documentation; and efficient communication to decision makers.

Criterion 4: Linkages and Feedback among Related Studies and Models

The environmental science work in the LOSLR Study, at least the contents presented for this review, use data and sub-models generated from the empirical wetland and SAR work in the IERM. The IERM, in turn, is used as a sub-model in the SVM as a means to present integrated environmental information. Linkages among the wetland, SAR, and IERM efforts are important for accurate interpretation and representation in the SVM and ultimately, in the regulation plans, themselves. Furthermore, linkages between the wetlands work and the FEPS model may also be important, as some of the physical, geomorphic properties of wetland habitat (such as dune- and barrier beach wetlands) can be strongly influenced by wave erosive action. The biophysical underpinnings needed to develop Performance Indicators for barrier beach wetland complexes are severely compromised by failure of the FEPS model to include accretion of dunes and protective effects of dunes and barrier islands on other parts of the shore. The development of a barrier beach Performance Indicator with linkages between the IERM and FEPS models appears imperative.

Linkages among SVM components could strengthen the integration and display of key information in the LOSLR Study. The FEPS model (see Chapter 3) developed scenarios to predict the response of the coastal lake bottom to various storms and outflow management options. These results might have been applied to the wetland studies. Alteration of bathymetry at the lower portion of wetlands may impact the rate of response of vegetation to water level change, not just in the lowest portion of a site but potentially in upper elevations if base level is altered. **Wetland researchers should collaborate with the shoreline modelers to develop bathymetric scenarios for wetland sites.**

The protected barrier beach wetlands of Lake Ontario comprise a major hydrogeomorphic community and are included in the wetland empirical work, but not integrated into the FEPS, IERM, or SVM models. A Performance Indicator exists for barrier beaches and sand dunes, but "no algorithm or economic calculations were developed for the FEPS due to inability of modeling sandy shore evolution over very long time periods" (Baird and Associates [no date], Barrier Beaches and Dunes, Performance Indicator Summary, Coastal Working Group). Evidence exists, however, that barrier beaches are being eroded away along the southern shore of Lake Ontario at Hamlin Beach State Park (Makarewicz et al., 2000; NYORHP, 1998), as armoring efforts are underway. The omission of barrier beach wetlands from the IERM and SVM could result in the detriment of an important hydrogeomorphic type and a NYSDEC protected wetlands along coastal Lake Ontario. **The LOSLR environmental science work depends on the ability of the FEPS model to provide the type of information needed by the SAR and IERM work, and the connections between FEPS and environmental work need strengthening or better documentation.**

Criterion 5: Spatial and Temporal Resolution and Scaling

The quarter month time step (see Chapter 2) presents challenges to achieve temporal resolution compatibility among different aspects of the LOSLR Study. Quarter month time steps were chosen for purposes of hydrologic/hydraulic models and the SVM (see Chapter 2), but the quarter month time step may be too coarse for biological applications. Temperature is a key stressor affecting fish reproduction in the St. Lawrence River, and it is considered as a stressor in LOSLR fish model(s). Fish models use a "daily" temperature time step for effects of water level and flow on spawning and early life histories. However in the IERM, there was a simplification of the IERM fish component, whereby a quarter month time step was introduced. Little or no information was provided on the rationale or impact of changing the time step from daily in the sub-model to quarter-monthly in the IERM. The accuracy of the IERM may be compromised by this discrepancy in time step. The type of explanation required to evaluate this scientific outcome would include:

- What is the daily variation in temperature compared to the quarter-monthly?
- Why did the fish population sub-model use a daily step but the IERM use a quarter-monthly step?
- What effect, if any, did this simplification of the model have on projected results?

Time scale resolution presents a similar concern for the use of quarter-monthly time steps for some bird populations. Rapid rise in water level is listed as an important hydrologic factor for some birds (IERM 1). Fan and Fay (2001, Variation of St. Lawrence River Hourly Water Levels about the Quarter-Monthly Mean, pages 8 and 9) point out that river level variation can be quite high during the spring freshet: "at Betancour, the daily level typically varies within ± 25cm from its quarter-monthly level, but can be as much as 80 cm higher or 50 cm lower year round." At Batiscan, daily levels can be 100cm higher or 70 cm lower than quarter monthly estimates. Drolet et al. (in IERM 1) report that eggs and chicks of the Virginia Rail are susceptible to drowning due to water level fluctuations. In fact, a rate of greater than 20 cm is suggested as being important, but the rate for this 20 cm rise is not noted (i.e., should have been represented as per day, per week, or per month). Assuming that the rate is 20 cm per day, the quarter-monthly time step is not acceptable because it is far too coarse. Similar situations are described for the Least Bittern and Black Tern. **Temporal scales need to be consistent between the empirical studies and models used in the IERM and subsequently in the SVM.**

Although the IERM 1 presented the 'typical wetland geometry profiles' (Figure 2-2, p. 22), there was no additional information to determine variability around the individual curves, and no attempt was made to validate these with an independent wetland subset from LO-USL. **Since wetland size can affect a wetland's vulnerability to perturbation and to a site's function as wildlife habitat, more attention must be paid to ensure that results for larger wetlands can be extrapolated directly to smaller ones in other geographic reaches of the Lake Ontario shoreline.**

Criterion 6: Thorough Documentation

Documentation presented serious problems for a thorough understanding, and therefore evaluation, of the wetlands, SAR, and IERM work. Some of the documentation problems stemmed from the review documents themselves. Draft reports, rather than completed reports, of much of the environmental sciences work were provided to the NRC committee, conveying that additional refinements would be forthcoming. Unfortunately, whole sections were missing from some of these draft documents. For example, in the IERM - 4 User's Manual entitled "The IERM Criteria Evaluation" was listed as not yet functional. Documentation was missing that clarified (1) who was responsible for the work; (2) what were the scope and plans for completing incomplete work; and (3) the timeline for work completion. At present, such calibration and documentation do not exist.

Specific examples of insufficient documentation on environmental sciences are given throughout this chapter. **Major points of concern focus around inadequate documentation of: (1) wetlands and SAR sampling choices; (2) choice of Performance Indicators in lieu of more traditional ecological assays; (3) methods to select and eliminate Performance Indicators in reducing the number from several hundred to 32; (4) resolution of temporal and spatial scales; and (5) treatment of error and uncertainty, and documentation should be provided for these areas of concern.**

Criterion 7: Effective Scientific Communication

Ecological Indexing

Indexing is a common practice in ecological sciences, used to present multivariate information in a simplified way. When done correctly and under appropriate conditions, indexing can be effective in conveying complex, intricate information. Still, indexing poses specific problems in the LOSLR Study. The Environmental Technical Working Group developed the single "Index of Ecological Integrity" (IJC, Draft Report), by collapsing the 32 key Performance Indicators into a single value that is a weighted average (group-weighted, region-weighted). As pointed out by Limno Tech (IERM 1), "it is important to note that the overall index of ecological integrity provides an overview of the key Performance Indicator results, *but the index by itself should not be considered sufficient to evaluate and rank plans* [italics added]." Rather, the entire suite of 32 key Performance Indicators should be used for comparison of plans in matrix and graphic formats such as the Target Diagram.

This review agrees that the Index of Ecological Integrity should not be used to evaluate potential regulation plans for at least two reasons. One reason is that the "Index of Ecological Integrity" is an oversimplification and has to be viewed with considerable caution. Forming an index assigns weights on the various components that go into the index, and these weights express priorities among Performance Indicators. The weights used to form the environmental index are hard to locate in the Board Room, and no justification is offered for differences in the weights that were assigned to regions, to groups of species or to individual species. The weighted priorities are elements of preference or value models, which vary among stakeholder groups.

Another reason is that a single index forces the same value model on all groups. Individual results of environmental Performance Indicators respond differently to variations in water regulation plans, and a single index will obscure these differences. Perhaps this concern is best illustrated by the example of the Moses Saunders dam. River segments, including the reservoir, upstream of the dam differ significantly from segments downstream of the dam. The problem with the simplified index is that a single value will be used to express conditions both above and below the dam. This example is stark, but this concern of obscuring important differences extends to plant and animal Performance Indicator values, as well. Furthermore, whether intended or not, it will be the tendency for non-scientists to use this single index value to evaluate all environmental issues. **For these reasons, the use of the Index of Ecological Integrity is not recommended for purposes of the LOSLR Study.**

OVERALL APPROPRIATENESS AND SUFFICIENCY OF ENVIRONMENTAL SCIENCE TO INFORM REGULATION PLAN OPTIONS

The third charge (Charge C) in the statement of task is aimed at determining whether the IJC documents presented for review are appropriate and sufficient for use in selecting a water regulation plan option. Three evaluation criteria are used to address this charge: the study scope and aims, balance between scientific and practical professional judgment, and future needs to respond to effects of water level regulation on multi-decadal time scale.

Criterion 8: Breadth of Study Scope

One of the impressive aspects of the LOSLR study is its inclusion of environmental considerations in selecting a water regulation plan. At its core, this research described plant species diversity and composition within vegetation zones for 32 wetland sites, upon which SAR and IERM Performance Indicators were developed. The documentation of the wetland and SAR species assemblages constitutes a significant database. Seldom is an analysis of so many sites possible in a single study. One lasting legacy of this project is the valuable database of standardized field measurements of many wetland components. In particular, the inventory of wetlands in Lake Ontario is a significant contribution, and should be preserved to support adaptive management and future research. It is important that GIS databases (such as those used for modeling different organisms) be maintained to ensure wide public accessibility and for possible enrichment as more data become available. Upon publication, this LOSLR study will contribute significantly to documentation of the relationship between flooding/dewatering history and wetland vegetation. **Every effort should be made to ensure that the data acquired for this project remain in public domain and available for future projects.**

Criterion 9: Balance Between Scientific and Practical Professional Approaches

When empirical data are lacking, it is sensible to make use of expert opinion to inform habitat models. Sometimes, though, expert opinions lack credibility when ad hoc methods are used to solicit them, rather than the more explicitly structured methods that have been developed in the field of decision analysis (e.g., Meyer and Booker, 1991; Morgan and Henrion, 1990). When used in standard practice, expert opinions can be invaluable complements or supplements to empirical data. Expert opinion was used in the SAR studies, but it lacked many of the standards of good practice for use, including (1) justification of choice of experts; (2) provision of a common background of available information to all experts; (3) orderly and itemized structure of expert judgments on related issues, rather than a composite judgment about population response; and (4) calibration of the experts' responses with whatever empirical data may be available. **These standards for expert analysis should be included in the SAR analyses and processes.**

Criterion 10: Identification of Future Needs

This body of environmental science, as presented in the wetlands, SAR, and IERM documents, is seen as a solid beginning to a science program aimed at understanding the LOSLR ecosystem. The data and information collected, analyzed and utilized in the LOSLR Study will provide a foundation for future, related scientific work that should be further developed, refined, expanded, and applied to scientific and policy uses.

As advances in science continue, the IERM and thus the SVM must be able to accommodate these advancements in science and modeling techniques. Currently, the IERM is a positive step forward for the management of lake levels and flows for the Lake Ontario/St. Lawrence system. In 2005, the LOSLR environmental field data are of short duration (less than three years) and spatially limited, and the propagation of error is essentially unknown. While the science is still under development, a common and sensible response to decision making hampered by limited information is adaptive management. Adaptive management can be designed to update and adapt the data and models used for decision making in a systematic way over time.

The environmental research explicitly focused on effects of water levels on wetlands, SAR, and other organisms in the LOSLR ecosystem. Other attributes of the aquatic system, such as water quality, shoreline urbanization, dredging and other sediment manipulation, were not included in these analyses. These other stressors likely have an influence on measures of environmental Performance Indicators, but their effects are undocumented in the LOSLR materials.

Because these other factors that are likely to influence SAR and other environmental Performance Indicators have not been included in any of the Performance Indicator models, it will be very difficult to use monitoring data to evaluate the success of water regulation plans and to update and improve the Performance Indicator models. There are likely to be variations in environmental status attributable to these or other factors in addition to or instead of water level fluctuations. **While the current design of the IERM and SVM cannot tease apart impacts caused by these factors from those due to water regulation, adaptive man-**

agement may provide an appropriate paradigm by which the SVM can be updated to include additional factors once the science becomes available.

SUMMARY ON ENVIRONMENTAL SCIENCES IN THE LOSLR STUDY

This chapter evaluated the three sectors of the LOSLR environmental sciences as a collective body of work. Findings and recommendations have been presented by Statement of Task Charges A, B, and C in the preceding sections. This summary presents major findings and recommendations separately for wetlands, SAR, and IERM, respectively, but first, issues are presented that are common to or span across the three sectors.

Cross-Cutting Findings in Environmental Sciences in the LOSLR Study

Improvements to the LOSLR environmental science work will take a range of effort and resources to address. Some choices made in the environmental science LOSLR studies are not optimal for the available information, but can be improved with a modest investment of effort. Examples of these include (1) calculating ratios (or another metric for comparing plans) in a consistent and coherent manner, (2) abandoning the Index of Ecological Integrity, (3) reformulating Performance Indicator models to remove any bias introduced by non-standard use of the precautionary principle; and (4) improved documentation for wetlands, SAR, and IERM work.

Other deficiencies can be remedied only with a more fundamental change in study approach, with a moderate investment in additional collection of field data and reworking, perhaps substantially reworking, LOSLR models. These include (1) defining dune-barrier Performance Indicators, (2) formulating a more ecosystem-oriented metric to represent at least some environmental Performance Indicators, (3) remedying the problems with temporal and spatial resolution, (4) reworking the FEPS model and its linkages to environmental models to facilitate better prediction of dune-barrier island responses and wetlands vegetation responses, (5) inclusion of SAR that had been excluded because they are not currently present in the study area, provided they meet other criteria for inclusion, and (6) use of appropriate methods for eliciting expert opinion.

Some issues can be remedied only over time and with considerable ongoing investment in data collection, analysis and integration with the decision making framework. These include (1) obtaining a more representative field sample of wetlands and aquatic vegetation, (2) incorporating factors other than water levels in data collection and modeling via an adaptive management scheme, and (3) validation and improvement of Performance Indicator models.

Wetland Evaluation Summary

In part or in whole, all 32 environmental Performance Indicators are based upon wetlands empirical data, which are used subsequently in IERM and SVM models. Therefore, the reliability of the wetland empirical data is directly related to the reliability of the SAR, IERM, and to SVM results. While applauded for the extensive database populated by the wetlands sampling inventories, the wetlands empirical work presents some concerns:

- The current sampling methods limited to shallow waters excluded or undersampled many of the submergent species in the high-quality wetlands and likely underestimated the diversity of submerged vegetation and available fish habitat
- The 32 "typical" wetland sites selected for sampling may not be representative of Lake Ontario-St. Lawrence River wetlands, as these sites were neither chosen randomly nor sampled to ensure representativeness of available wetland types. Therefore, the validity of applying study results to the entire wetland inventory remains in question
- Wetlands results need to be validated using data on those or similar wetlands from previous studies (i.e., IJC Water Levels Reference Study, Wilcox et al., 1992) to demonstrate the quality of the analysis and the certainty of wetland forecasts
- Propagation of error is a major concern for the LOSLR. There is a potential for error to be propagated from the wetlands empirical data through the SVM, but this potential is not documented and therefore cannot be quantified
- Specific attention is needed to ensure that results for larger wetlands can be extrapolated directly to smaller ones in other geographic reaches of the LOSLR system
- Wetland data acquired in this effort should remain in public domain and preserved to support adaptive management and future research

SAR Evaluation Summary

For many SAR, the Performance Indicator models developed during the LOSLR studies may be the best that can be done with the data that could be gathered and analyzed with the time and resources made available during the study. Documentation was problematic in SAR materials, and improved documentation is needed for:

- use of expert opinion;
- selection of SAR for analysis; and
- criteria used to evaluate confidence in the models and sensitivity of the Performance Indicators to water level regulation.

IERM Evaluation Summary

The IERM and SVM were developed to answer the question: Can the impact of regulating Lake Ontario lake levels and St. Lawrence River flows be mitigated to improve environmental impacts compared to Plan 1958DD? From an environmental perspective, the IERM is a single-species approach—not an ecosystem approach—that does not include other stressors on abundance and performance of organisms. As a result, with time, the IERM will become less accurate as a predictor of performance. The main findings about and recommendations for improving the IERM follow:

- Performance Indicator ratios are calculated differently, based on different units, such as biomass, area, and other measures, and should be reconciled to calculate ratios in a consistent, comparable, and coherent manner
- An explanation is needed about how Performance Indicators were selected or eliminated for use in the IERM and SVM
- Better documentation is needed about how limitations of certainty and propagation of error are calculated, including clarification of or a quantitative substitute for the 10 percent rule that was derived from a Study Board's "administrative decision"
- The Performance Indicator for barrier beaches should be included in the IERM
- The Index of Ecological Integrity used in the LOSLR Study presents a single value that (1) obscures differences among environmental Performance Indicators and (2) simultaneously attempts to represent conditions upstream and downstream of the Moses-Saunders dam. It is recommended that the index not be used for the LOSLR Study
- The quarter month time-step of the SVM needs to be reconciled with the shorter time steps used in the SAR and faunal sub-models in the IERM
- Calibration, validation and error propagation must be estimated or better documented to increase confidence in IERM results.

Certainly, the IERM model produced is better than the 1958DD Plan of Operation model previously used for making decisions, but the IERM may not be adequate as a predictive management tool because of issues of validation, calibration, cumulative uncertainties, omission of critical Performance Indicators, the lack of an ecosystem approach and a resulting obsolescence.

***

In sum, this body of environmental science, as presented in the wetlands, SAR, and IERM documents, is seen as a satisfactory beginning to a long-term program for understanding broad impacts of water regulation in the LOSLR system. The data and information collected will provide a solid foundation for future, related scientific work and policy decisions for Lake Ontario and the St. Lawrence River. Currently, this environmental work is compromised by unanswered questions in the wetlands sampling methodologies and inadequate documentation that ultimately undermines confidence that linkages among environmental Performance Indicators, IERM model and its sub-models, and the SVM operate as designed. Several issues need to be addressed in order to elevate the environmental work to the level of being appropriate and sufficient to inform the water regulation plan decision making process. Propagation of error and the overall thin treatment of uncertainty need to be addressed with rigor; the calculation of Performance Indicator ratios needs to be explained; and criteria and methods for the selection or elimination of Performance Indicators needs documentation. Even still, the breadth of the environmental sciences in the LOLSR Study and the empirical databases created from these efforts are a step in the right direction and commended. It is recommended that the wetlands, SAR, and the IERM work become the foundation for an adaptive management program whereby the environmental Performance Indicators and SVM can be adjusted appropriately as the science of the LOSLR system advances.

REFERENCES

Ashworth, S. M. 1997. Comparison between restored and reference sedge meadow wetlands in south-central Wisconsin. Wetlands 17:518-527.

Baldwin, A. H., and I. A. Mendelssohn. 1998. Effects of salinity and water level on coastal marshes: an experimental test of disturbance as a catalyst for vegetation change. Aquatic Botany 61:255-268.

Bedford, K. W. 1992. The physical effects of the Great Lakes on tributaries and wetlands. J. Great Lakes Research 18:571-589.

Campbell, J. M. 1992. Habitat and population ecology of Lobelia Kalmii on Presque Ilse, PA correlated with changes in Lake Erie water levels. J. Pennsylvania Academy of Science 66:123-127.

Chow-Fraser, P. 2005. Ecosystem response to changes in water level of Lake Ontario marshes: lessons from the restoration of Cootes Paradise Marsh. Hydrobiologia. 539:189-204.

Chow-Fraser, P., V. L. Lougheed, B. Crosbie, V. LeThiec, L. Simser, and J. Lord. 1998. Long-term response of the biotic community to fluctuating water levels and changes in water quality in Cootes Paradise Marsh, a degraded coastal wetland of L. Ontario. Wetland Ecol. Manage. 6: 19-42.

Dieffenbacher-Krall, A. C. 1996. Paoleo- and historical-ecology of the Cutler grasslands, Cutler, Maine: implications for future management. Natural Areas Journal 16:3-13.

Eyles, N., M. Doughty, J. I. Boyce, M. Meriano, and P. Chow-Fraser. 2003. Geophysical and sedimentological assessment of urban impacts in a Lake Ontario watershed and lagoon: Frenchman's Bay, Pickering, Ontario. Geoscience Canada. 30:115-128.

Fan, Y., and D. Fay. 2001. Variation of St. Lawrence River Hourly Water Levels About the Quarter-Monthly Mean. Hydrology and Hydraulics Technical Working Group. Lake Ontario–St. Lawrence River Study, International Joint Commission. December 2001.

Farney, R. A., and T. A. Bookhout. 1982. Vegetation changes in a Lake Erie marsh (Winous Point, Ottawa County, Ohio) during high water years. Ohio Journal of Science 82:103-106.

Franklin, J. F. 1993. Preserving biodiversity: species, ecosystems or landscapes? Ecological Applications 3(2):202-205.

Gaudet, C. L., and P. A. Keddy. 1988. A comparative approach to predicting competitive ability from plant traits. Nature 334(6179):242-243.

Grace, J. B. 1988. The effects of nutrient additions on mixtures of *Typha latifolia L.* and *Typha domingensis Pers.* along a water-depth gradient. Aquatic Botany 31:83-92.

Grace, J. B., and R. G. Wetzel. 1998. Long-term dynamics of *Typha* populations. Aquatic Botany 61: 137-146.

Greenway, K. J., and V. J. Lieffers. 1997. A boreal forest grass with an open meadow photosynthetic strategy. Canadian J. Botany 75:562-567.

Howard, R. J., and I. A. Mendelssohn. 1995. Effect of increased water depth on growth of a common perennial freshwater–intermediate marsh species in coastal Louisiana. Wetlands 15:82-91.

Hupp, C. R. 2000. Hydrology, geomorphology, and vegetation of Coastal Plain rivers in southeastern USA. Hydrological Processes 14:2991-3010.

IERM 1. Limno-Tech, Inc. DRAFT of April 2005. Development of an Integrated Ecological Response Model (IERM) for the Lake Ontario-St. Lawrence River Study.

IERM 2. IERM Model (version 4.1.1).

IERM 3. Limno-Tech, Inc. DRAFT of March 2004. Development of the_Integrated Ecological Response Model (IERM) for the Lake Ontario-St. Lawrence River Study.

IERM 4. Limno-Tech, Inc. April, 2005. Integrated Ecological Response User's Manual.

IERM 5 RS. Turgin, K, O. Champoux, S. Martin, and J. Morin. April, 2005. Modelisation des grandes classes de milieux humides de la plaine inondable du fleuve Saint-Laurent: Consideration de la succession des communautes vegetales.107 Wetlands Model for the Lower St. Lawrence River [French].

IERM 6 RS. Morin, J., O. Champoux, S. Martin, and K. Turgeon. April, 2005. Modelisation integree de la reponse de l'ecosyteme dans le fleuve Saint-Laurent: Rapport final des activities enterprises dans le cadre du Plan d'etude sur la regularisation du lac Ontario et du fleuve Saint Laurent. 108 Species at Risk Model for the Lower St. Lawrence River [French].

Kao, J. T., J. E. Titus, and W. Zhu. 2003. Differential nitrogen and phosphorus retention by five wetland plant species. Wetlands. 23: 979-987.

Keddy, P. A. 1984. Quantifying a within-lake gradient of wave energy in Gillfillan Lake, Nova Scotia. Canadian J. Botany 62:301-309.

Kelley, J. C., T. M. Burton, and W. R. Enslin. 1984. The effects of natural water level fluctuations on N and P cycling in a Great Lakes marsh. Wetlands 4:159-175.

Keough, J. R., T. A. Thompson, G. R. Guntenspergen, and D. A. Wilcox. 1999. Hydrogeomorphic factors and ecosystem responses in coastal wetlands of the Great Lakes. Wetlands 19:821-834.

Kercher, S. M., Q. J. Carpenter, and J. B. Zedler. 2004. Interrelationships of hydrologic disturbance, reed canary grass (Phalaris arundinacea L.) and native plants in Wisconsin wet meadows. Natural Areas Journal 24: 316-325.

Kozlowski, T. T. 1984. Plant responses to flooding of soil. Bioscience 34:162-167.

Lauck, T., C. W. Clark, M. Mangel, and G. R. Munro. 1998. Implementing the precautionary principle in fisheries management through marine reserves. Ecological Applications 8(1) Supplement: Ecosystem Management for Sustainable Marine Fisheries S72-S78.

Lenssen, J., F. Menting, W. Van Der Putten, and K. Blom. 1999. Control of plant species richness and zonation of functional groups along a freshwater flooding gradient. Oikos 86: 523-534.

Lessman, J. M., I. A. Mendelssohn, M. W. Hester, and K. L. McKee. 1997. Population variation in growth response to flooding of three marsh grasses. Ecological Engineering 8:31-47.

Li, S., S. R. Pezeshki, and S. Goodwin. 2004. Effects of soil moisture regimes on photosynthesis and growth in cattail (*Typha latifolia*). Acta Oecologica. 25: 17-22.

Lougheed, V. L., T. Theÿsmeÿer, T. Smith, and P. Chow-Fraser. 2004. Carp exclusion, food-web interactions, and the restoration of Cootes Paradise Marsh. J. Great Lakes Res. 30 (1): 44-57.

MacDonald, S. E., and V. J. Lieffers. 1993. Phizome plasticity and clonal foraging of Calamagrostis Canadensis in response to habitat heterogeneity. Journal of Ecology 81:769-776.

Makarewicz, J. C., J. M. Haynes, R. C. Dilcher, C. J. Norment, and T. Lewis. 2000. Biological Survey of Yanty Creek Marsh. New York Office of Parks and Recreation and Historic Preservation. Castile, NY: Western District, Genesee Region, One Letchworth State Park.

McKee K. L., and I. A. Mendelssohn. 1989. Response of a freshwater marsh plant community to increased salinity and increased water-level. Aquatic Botany 34:301-316.

McMaster, R. 1997. Floristics, zonation and succession of vascular vegetation in fifteen beaver-impacted wetlands in western Massachusetts. Dissertation. University of Massachusetts.

Mendelssohn, I. A., and D. M. Burdick. 1988. The relationship of soil parameters and root metabolism to primary production in periodically inundated soils. P. 398-428. IN D. D. Hook (eds). The Ecology and Management of Wetlands. Vol. 1. Ecology of Wetlands. Portland OR: Timber Press.

Meyer, M., and J. Booker. 1991. Eliciting and Analyzing Expert Judgment, a Practical Guide. London, UK: Academic Press.

Middleton, B. 2002. Nonequilibrium dynamics of sedge meadows grazed by cattle in southern Wisconsin. Plant Ecology 161: 89-110.

Morgan, M., and M. Henrion. 1990. Uncertainty. A Guide To Dealing With Uncertainty in Qualitative Risk and Policy Analysis. New York: Cambridge University Press.

Mulhouse, J. M., and S. M. Galatowitsch. 2003. Revegetation of prairie pothole wetlands in the mid-continental U. S.: Twelve years of post-reflooding. Plant Ecology 169:143-159.

Naiman, R. J., H. Décamps, and M. Pollock. 1993. The role of riparian corridors in maintaining regional biodiversity. Ecological Applications 67:1254-1269.

NYORHP (New York Office of Parks, Recreation and Historic Preservation). 1998. State Environmental Quality Review. Provide Shoreline Protection–Yanty Marsh, Hamlin Beach State Park.

Ovenden, L., and G. R. Brassard. 1989. Wetland vegetation near Old Crow, Northern Yukon. Canadian J. Botany 67:954-960.

Pollock, M. M., R. J. Naiman, and T. A. Hanley. 1998. Plant species richness in riparian wetlands—A test of biodiversity theory. Ecology 79(1):94-105.

Robert, M., P. Laporte, and R. Benoit. 2000. Summer habitat of yellow rails, Cotunicops noveboracensis, along the St. Lawrence River, Quebec. Canadian Field-Naturalist 114:628-635.

SAR 1. Water Fluctuation Impacts on Species At Risk Cornwall to Pointe due Lac [French].

SAR 2. Lake Ontario Species at Risk Supplement (Least Bittern and Black Tern Reproductive Index PIs).

SAR 3A. Impact of Water Level Regulation on Nearshore Habitat Availability and SAR.

SAR 3B. (Supplement) Impact on dunes and SAR.

SAR 4. Species at Risk Fish Supplement (Year 4 Modeling Group Report).

SAR 5. Lower St. Lawrence Species at Risk Final Report (2005) [French].

Sorrell, B. K., I. A. Mendelssohn, K. L. McKee, and R. A. Woods. 2000. Ecophysiology of wetland plant roots: A modeling comparison of aeration in relation to species distribution. Annals of Botany 86:675-685.

Tessier, C., A. Aubin, and C. Chenevier. 1984. Elevation and distance factors in the structure and productivity of a riverine macrophyte community. Canadian J. Botany 62:2260-2266.

Wei, A., and P. Chow-Fraser. 2005. Untangling the confounding effects of urbanization and high water level on the cover of emergent vegetation in Cootes Paradise Marsh, a degraded coastal wetland of Lake Ontario. Hydrobiologia. 544: 1-9.

Wei, A., and P. Chow-Fraser. 2006. Synergistic impact of water-level fluctuation and invasion of *Glyceria* on *Typha* in a freshwater marsh of Lake Ontario. Aquat. Bot. (In press).

Wei, A., P. Chow-Fraser, and D. Albert. 2004. Influence of shoreline features on fish distribution in the Laurentian Great Lakes. Can. J. Fish. Aquat. Sci. 61: 1113-1123.

Wilcox, D. A., J. E. Meeker, and J. Elias. 1993. Impacts of water-level regulation on wetlands of the Great Lakes. Phase 2 Report to Working Committee 2. International Joint Commission Water-levels Reference Study, Ottawa, Ontario. Canada and Washington, DC. USA.

Wilcox, D. A., S. I. Apfelbaum, and R. D. Hiebert. 1984. Cattail invasion of sedge meadows following hydrologic disturbance in the cowls Bog Wetland Complex, Indiana Dunes National Lakeshore. Wetlands 4:115-128.

W 1. Wilcox, D. A., K. P. Kowalski, and M. L. Carlson. 2003. Year 3 Report of Relationship between Lake Ontario Water Levels and the Wetlands. Ann Arbor, MI: Great Lakes Science Center, U.S. Geological Survey.

W 2. Evaluation of Water Level Regulation Influences on Lake Ontario and upper St. Lawrence River Coastal Wetland Plant Communities.

Chapter 5
Toward Adaptive Management

This chapter draws together findings and recommendations from the NRC/RSC review of LOSLR studies and models. It begins by situating those studies and models within the historical context that shaped them, follows with the Statement of Task, noting the five qualifications of this review, and concludes with five overarching conclusions and recommendations. The conclusions look back at what has, and has not, been accomplished; while the recommendations look forward toward continuing scientific approaches needed to address the identified gaps, deficiencies, and water level and flow regulation challenges.

HISTORICAL PERSPECTIVE

The historical context of the LOSLR studies and models includes a half-century of experience with water regulation plans and their effects (Clinton Edmonds and Associates, 2002). The first regulation plan for the Lake Ontario-St. Lawrence system (Plan 12-A-9) was adopted in 1956 (International Joint Commission, 1956) and replaced in 1960. The second Plan 1958-A was replaced after two years, and the third Plan 1958-B after only one year. The current Plan 1958DD was adopted in 1963 and has been in effect for 42 years, but it has required an increasing number and duration of deviations to accommodate hydroclimatic, ecological, and socioeconomic changes. In the 1990s, alternative plans were formulated to improve operations and reduce deviations, but they were not adopted in large measure because they did not examine the broad range of water regulation impacts or offer sufficient improvement.

The IJC responded to these deficiencies by commissioning the LOSLR *Plan of Study* (BACK 2). The LOSLR Study was authorized to expand the scope of inquiry to include environmental, social, and economic impacts. To achieve that aim, the LOSLR Study developed a Shared Vision Model approach to formulate and evaluate RPOs through a collaborative process with nine Technical Work Groups and a Public Interest Advisory Group. The LOSLR Study thus sought to address, in five years, water regulation issues that have developed over a half-century. These water regulation issues will continue to evolve in future decades in response to new socioeconomic, ecological, and hydroclimatic challenges. The LOSLR Study addresses some of these future challenges quantitatively (e.g., climate change scenarios, which were not part of this review), and others qualitatively (e.g., through Conceptual Narratives).

SCOPE AND LIMITS OF THIS REVIEW

The IJC requested the National Research Council and Royal Society of Canada, "...to evaluate the appropriateness and sufficiency of the studies and models used to inform decisions related to regulation plan options." The Statement of Task reads:

> The Lake Ontario/St. Lawrence River program science, as represented in the reports and model documentation provided, will be reviewed by in terms of the degree to which:
>
> (1) The studies reflect reasonable scientific methods, assumptions and supported findings;
> (2) The models sufficiently and appropriately integrate and display the key information needed for a comprehensive evaluation and understanding of the tradeoffs for selecting among the candidate RPOs; and
> (3) The models and reports are sufficient and appropriate to evaluate the various regulation plan options (RPOs) and impacts of changes in water levels and flows.

There are five qualifications to keep in mind when using this review:

- **In some cases interim documents were presented for review.** The documents presented for review were in various stages of completion. In cases where documents were incomplete, the committee tried to procure the most current version of the work. The committee treated the documents presented by the IJC as representative of the science under review and recognized that some documents would be modified after the review.
- **This is a selective review of LOSLR studies and models.** The committee reviewed selected documents for the SVM, FEPS, and environmental sciences. The review does not encompass all of the LOSLR research undertaken in these three fields. Nor does it include scientific fields in the LOSLR study that lie outside the scope of review but have some relevance for the overall charge, such as hydrology and hydraulics, navigation, hydropower, M&I, RSPM, etc. To the extent that the Shared Vision Model incorporates results from these other fields, this review offers a partial perspective on the overall sufficiency of LOSLR studies and models.
- **The review occurs toward the end of the 5-year LOSLR Study.** The NRC review was initiated in the final year of the LOSLR Study. This timing offers an opportunity to evaluate the studies and models used to inform to decision makers, as well as to identify opportunities for improvement prior to submission. However, some recommendations would have been more useful at the beginning or middle of the study period rather than this close to the Study's completion.
- **The review concentrates on the science for evaluating water level and flow effects of RPOs and for informing decision makers, and not on the**

RPOs themselves or on decision making policies. Ten scientific evaluation criteria were used to evaluate the LOSLR studies and models. These criteria are common to the scientific and practical professional disciplines involved in evaluating complex studies, such as the water level and flow effects of regulation plan options in the Lake Ontario-St. Lawrence River system.

- **The review distinguishes among conclusions and recommendations in terms of their certainty, importance, and ability to fix deficiencies.** The conclusions of this report vary in terms of their certainty due to the state-of-the-science in different fields and gaps in study documentation. Some conclusions have more importance to the success of the LOSLR Study than others. Points of study weakness and recommendations vary in the degree to which they can be fixed and the amount of time and additional research needed to address them. The review strives to distinguish among recommendations that entail short- and long-term action. Depending upon the level of commitment made, short-term conclusions and recommendations are those that can be addressed in a period of weeks to months without new research. At this stage of the process, substantive research deficiencies cannot be fixed. Thus, short-term recommendations concentrate on improving the documentation, scientific communication, and disclosure of potential implications of these limitations for decision makers. Longer-term recommendations require investment in additional data collection, analysis, and interpretation. In making these distinctions, conclusions deemed to have the most certainty, importance, and timeliness are presented as declarative statements. Conclusions that have less certainty, importance, or timeliness are presented in conditional terms. Conclusions that have the least certainty, importance, or ability to be fixed are raised as questions.

With these five limitations in mind, this chapter presents five overarching findings about the LOSLR studies and model accomplishments, their scientific strengths and weaknesses, and the recommendations that emerge from this review.

EVALUATION

This review commends some aspects of the studies and models and raises substantial concerns about others. If all of the individual reports, studies, and models reviewed here were found to be appropriate and sufficient, the committee might come to a qualified affirmation of their collective sufficiency for informing decision making. Instead, while commending advances over previous studies, this review finds deficiencies in LOSLR studies and models for each of the ten evaluation criteria. The overarching conclusions are:

1. **LOSLR studies and models expand interdisciplinary scientific inquiry on the potential environmental effects of water level and flow regulation options in the Lake Ontario-St. Lawrence River Basin in ways that are useful for informing decision making in some respects**. The LOSLR studies undertook a broad scope of inquiry and participatory process for understanding the potential effects of water level and flow regulation in a complex water system. The

Shared Vision Model compiles effects of flow regulation on environmental, coastal, and recreational effects, in addition to statutory obligations to consider municipal, navigation, and hydropower uses. LOSLR models go beyond previous Great Lakes water regulation efforts in compiling results of scientific inquiry and stakeholder input. Its iterative public participation process begins to treat water level and flow regulation as an adaptive process. Identification and inclusion of environmental performance indicators is a major contribution. The LOSLR studies have created large new databases, like those on wetland vegetation and coastal land use—that did not exist previously and that could, if archived and made readily accessible, have continuing value. However, the studies and models need to go further in encompassing ecological and socioeconomic scenarios and the linkages among them. Moreover, while the studies and models attempt to balance scientific and practical professional water management approaches, the findings below indicate where established standards of inquiry were, and were not, met.

2. **The studies and models vary widely in empirical support, and they need stronger and more consistent quality control, quality assurance and treatment of error and uncertainty needed to inform decision making.** This review finds that each of the studies has scientific merits and deficiencies that are summarized here. Following the framework used throughout the review, the scientific evaluation criteria are discussed under the headings of empirical foundations, quality assurance and quality control, and treatment of error and uncertainty.

- **Empirical foundations (e.g., data, sampling, analysis).** The studies and models reviewed here focus primarily on environmental and coastal investigations. The reviewed materials have varied types of data, data quality, sampling methods, and aggregation techniques that pose challenges for using results to inform management decisions. For example: (1) the coastal research developed a detailed land use parcel database, but that database differs in coverage for Canada and the USA; (2) although detailed for the sites selected, wetlands sampling is not statistically representative of wetland vegetation types along the coast or of deeper submergent vegetation in deeper water; (3) species at risk (SAR) data were collected to help develop some but not all SAR performance indicators; and (4) questions were raised about the way that some regression models were used. Empirical variation is to be expected in a multidisciplinary study where each discipline brings its own conventions, and in which long-term data collection is limited by the five-year timescale. Still, differences in empirical methods can propagate error and uncertainty in ways that have relevance for decision making. **Recommendation: As no new data can be collected in the near-term, LOSLR study final reports should underscore empirical limitations, data gaps, and sampling problems, and discuss their implications for decision making.** [1] **For the longer-term,**

---

[1] Lake Ontario-St. Lawrence River Study Board (LOSLR). 2005. Options for Managing Lake Ontario and St. Lawrence River Water Levels & Flows: Final Report. July 1, 2005 draft; on file at NRC-WSTB.

research to correct deficiencies in data and models including replacement of regression equations with process models should be prioritized.

- **Quality assurance and quality control (e.g., model validation, verification, and calibration; use of expert judgment; and independent peer review).** The models generally lack adequate validation, verification, and calibration. In some cases, validation may have occurred or is briefly mentioned but is not documented. In other cases, it appears not to have been undertaken. For example, the SVM had not yet received its "stamps of approval" confirming model validation from technical work group members. Reports on the FEPS model suggest that model calibration has occurred but do not provide detailed documentation. Validating applications of proprietary models such as COSMOS is important for full scientific peer review. Environmental studies lack consistent protocols for quality assurance, and the IERM model acknowledges that validation was not attempted. The SVM and FEPS models and SAR studies make creative use of expert judgment, which should be subject to more formal quality assurance by using well-established protocols for eliciting expert judgments. Some studies rely on "peer review" by fellow team members while others involve refereed papers. Earlier external scientific review of the overall LOSLR study program would have been timely for identifying and rectifying deficiencies. **Recommendation: In the short-term, LOSLR final reports should inform decision makers of the types of quality assurance measures that were, and were not, undertaken and discuss their potential implications for decision making. Further independent scientific review of final reports is recommended. In the longer-term, rigorous quality assurance methods should be put in place for evaluating the effects of water level and flow regulation.**

- **Treatment of error and uncertainty.** This criterion was not fulfilled in the studies or models reviewed at the level expected for informing decision making. The SVM, FEPS, and IERM models do not present an overall framework for uncertainty analysis, which should include natural variability, data uncertainties, model uncertainties, model parameter uncertainties, and decision model uncertainties. Some individual studies, such as wetlands vegetation analysis, address natural variability and indicate error bars. The SAR 3A report provides a good model for qualitative discussion of uncertainty. The Shared Vision Model treats the uncertainty of environmental indicators with a simplistic, and unexplained, 10% criterion, and it does not apply any uncertainty estimate to economic indicators. Linkages among LOSLR studies and models lead inherently to the propagation of uncertainties, but SVM documentation does not analyze those cumulative uncertainties or discuss their implications for informing decision making. Without formal analysis and discussion, it is not possible to assess the types or magnitudes of error and uncertainty for particular water regulation plans, or to know whether differences between plans are significant. **Recommendation**: **In the short-term, LOSLR final study reports should inform decision makers of the uncertainties that were analyzed, those that were not analyzed, and their potential implications for de-**

cision making. Future studies of water level regulation effects in the LOSLR basin should develop a comprehensive approach to uncertainty analysis.

3. The LOSLR models and studies reviewed here do not adequately integrate and display the key information needed for comprehensive evaluation and understanding of the tradeoffs for selecting among the candidate regulation plans. This conclusion is based on the following four review criteria:

- **Linkages and feedbacks among related studies and models.** "Comprehensive evaluation and understanding of tradeoffs among RPO alternatives" requires a system dynamics approach that models the linkages and feedbacks among socioeconomic and environmental processes. The SVM compiles first-order effects on environmental, coastal, and other indicators generated by FEPS, IERM, and other models. But, as the IERM user's manual indicates (IERM 4), it is not an ecosystem model that incorporates the feedback effects of water level variation on species and habitat conditions. Instead, it compiles initial impacts (first-order effects) on performance indicators, and it is thus an impact accounting model rather than an ecosystem model. In terms of model linkages, the FEPS model alters the bathymetry of shoreline environments, but those bathymetric changes were not fed into the IERM to alter the extent or depth of wetland inundation, which could in turn affect the vegetation, shoreline habitats, and other environmental performance indicators associated with water level variation. These vegetation changes could have feedback effects on sediment transport and coastal erosion. External model linkages include economic and demographic scenarios that are relevant for evaluating candidate water regulation plans to replace Plan 1958DD. For example, real estate values of coastal property continue to rise at rapid rates, and the demand for different water and related land uses is changing, but the SVM does not incorporate such scenarios in its structure.

This report acknowledges that some of these linkages and feedbacks require knowledge beyond the current limits, and that fact should be discussed in the final reports and presentation of SVM results. However, other linkages and feedbacks between the SVM and its sub-models, and externally between the SVM and scenarios of socioeconomic change, could have been addressed. The reviewed studies and models make progress toward comparing the effects of regulation plan options, but the comparisons reviewed here do not provide a *comprehensive* basis for evaluating and understanding tradeoffs among regulation plan options. **Recommendation: In the short-term, the LOSLR final reports should inform decision makers of what has, and has not yet, been accomplished in the way of integrated water and environmental systems modeling. As part of an ongoing program, a LOSLR modeling system that dynamically links and reflects feedback among sub-models is recommended.**

- **Treatment of spatial and temporal resolution and scaling.** Scaling issues in the Lake Ontario-St. Lawrence River basin are challenging. The LOSLR studies involve a wide range of spatial and temporal scales, which raise a number of concerns. For example, although more detailed hydrologic time series and station data are available for use in the STELLA model in the SVM, the model generates a single series of quarter-monthly values for the level of Lake Ontario, based on historical water management practice. Use of these single series values can result in a loss of precision, as the quarter-month does not provide enough temporal variation for many environmental impacts for fish, SAR, and wetlands. This coarse time step was recognized as a potential problem in the LOSLR *Plan of Study*, which called for a 2D hydrodynamic model for the St. Lawrence River that operated on fine enough time scales to supplement the quarter-monthly time step generated by the SVM. As noted earlier, the LOSLR approach of using quarter-monthly values in Lake Ontario to calculate water levels for selected stations in the upper St. Lawrence River through regression analysis is inferior to hydrodynamic flow routing, and the combined use of regression and hydrodynamic models needs to be more fully explained. The FEPS model uses the single lake level elevations along with a grid of wind and wave fields that ultimately erode and flood individual shoreline parcels and reaches, the results of which are aggregated back to lake-wide effects. The errors and uncertainties associated with these varied resolutions and scales of inquiry need fuller analysis and discussion, as errors may exceed the differences between model outputs for some performance indicators and plans. **Recommendation: In the short-term, the LOSLR final reports must inform decision makers of temporal and spatial scaling issues that affect the accuracy and uncertainty of predictions of regulation effects. In the longer term, choice of time step should better reflect the critical response times for system indicators, including those where transient fluctuations in water temperature and water level are critical, and appropriate hydraulic and hydrodynamic modeling approaches should be implemented.**

- **Thorough documentation of scientific studies and models.** Of the ten criteria employed in this review, inadequate documentation was the most frequently cited deficiency. It is one of the few deficiencies that can be corrected in the near term. The Shared Vision Model had the least amount of documentation presented for this review, and the documentation that was presented was not at a level of completion ready for external scientific and public review, although it may have been adequate for internal working by the Study Board. FEPS included more detailed descriptions of modeled performance indicators than other studies, but did not fully document the models themselves. A user's manual exists for the IERM and provides partial documentation, but insufficient explanations of weighting and aggregation in the model. Exceptions to these general patterns include the SAR 3A and 3B reports, which are well documented. Better documentation might have addressed some of the scientific concerns raised in this review, and might have raised additional questions. In either case, fuller disclosure of what was done and how it was done is needed to inform decision making.

Because the SVM is the primary tool for understanding and evaluating tradeoffs among RPOs, we concentrate on it in this recommendation. Documentation of the SVM should include more complete discussion of its role in the Shared Vision planning process. It should describe SVM development and refinement, including standard technical documentation of all component models. It should describe how scientific and stakeholder criteria were interactively used to formulate, screen, and evaluate the range of choice among RPOs. **Recommendation: In the short-term, LOSLR final reports should include a thorough documentation of studies and models, especially the Shared Vision Model, and seek further independent scientific review of those reports.**

- **Effective scientific communication.** This criterion involves the presentation of scientific information in ways that ensure its reception and comprehension by scientific, public, and decision making groups. Effective communication addresses public interests, communicates scientific findings and uncertainties clearly, is received accurately, and supports decision making.

The efficacy of scientific communication varies among LOSLR studies and models. Information about the environmental studies and performance indicator summaries were conveyed more clearly than the sub-model documentation, and sub-model documents more clearly than the SVM documentation. An example of deficient, or even misleading, communication is the differential treatment of economic and environmental indicators in Board Room presentations. On the other hand, the SVM Board Room spreadsheet tables and graphics have strong potential as a vehicle of scientific communication. As indicated in chapter 2, using the Board Room with broader public groups would require some editing and graphic design improvements. **Recommendation: In the short-term the LOSLR final reports should communicate their scientific results with transparency to support decision making and to give full treatment of uncertainties and non-scientific dimensions of the studies. In the longer-term, the SVM Board Room may be refined for continued use as a vehicle for scientific communication.**

4. **Despite the breadth of LOSLR studies and models, ongoing analysis is needed to provide a strong scientific basis for *long-term* decision making about water level and flow regulation in the Lake Ontario-St. Lawrence River basin.** Three points support this conclusion. First, current knowledge about the lower Great Lakes system is not comprehensive. The LOSLR studies and models broaden understanding about potential effects of RPOs, but a more comprehensive modeling approach is needed to understand system feedbacks, linkages, and uncertainties. A system dynamics model would: (a) improve the physical system description; (b) identify the most important feedback relationships; and (c) improve understanding of feedback effects on system behavior. Some feedback relationships would require expansion of model boundaries so that key processes, ranging from coastal urbanization and regional economic growth to climate change, are incorporated and their impacts are made visible within the model.

Second, previously selected regulation plans have been periodically reviewed and replaced in light of new knowledge. Initial RPOs were replaced due to unanticipated hydrologic and hydraulic events and their consequences. Plan 1958D accommodated some of those phenomena but required deviations to address less well-understood and emerging socioeconomic and environmental values, such as recreational boating and wetlands protection. The LOSLR studies and models begin to address those effects. However, any plan adopted on the basis of current science without provision for regular updating as knowledge advances is likely to prove inadequate within a timeframe of years to decades.

Third, the LOSLR models evaluate effects of *future* RPOs and hydrologic scenarios on *historical* and *current* environmental and social performance indicators. This is important, given the significance of hydroclimatic variability for RPOs and the challenges of modeling current environmental and socioeconomic processes. Although this report does not review the climate change research and scenarios, it commends the inclusion of global processes that affect the robustness of regional RPO decisions (Sousounis, 2002). In the future, however, RPO decisions will require comparable scenario development and evaluation for other environmental and social processes. Changes in regional economic structure, demography, water demand, transportation technology, coastal land use, and socioeconomic values—to name a few—will likely transform the profile of stakeholder interests, performance indicators, and socioeconomic impacts associated with RPOs (Economic Advisory Committee, 2004). The past half-century indicates that these types of structural shifts in socioeconomic and environmental conditions and values, in conjunction with hydrologic variability, have had substantial implications for RPO decision making.

The LOSLR studies and models begin to address these issues through brief conceptual narratives with a planning horizon of 10 to 15 years that are linked to the Shared Vision Model, as recommended by the Economic Advisory Committee (2004). The conceptual narratives employ a common template, but they vary in detail, completeness, and level of peer review (Board Room, FEPS 11-13; and Leger, 2005). Correcting the scientific and modeling deficiencies identified in this review is necessary and appropriate, but not sufficient for informing water regulation decisions on a multi-decadal timescale. **Recommendation: In the short-term, the LOSLR Study should complete the conceptual narratives and external peer review. For the longer-term, the IJC should consider commissioning an ongoing management and monitoring system to feed the results of current choices for water level regulation back into a dynamic model of the LOSLR system, in order to improve the scientific basis for future planning on a multidecadal timescale.**

5. Build upon the LOSLR studies and models through an Adaptive Management Program. The LOSLR Study may begin a new approach to water level and flow regulation in the Great Lakes. Even when deficiencies are corrected, there will be an inherent need for continuing scientific study and modeling of the actual effects of present RPO decisions, as well as the possible effects of future RPO adjustments. These needs are anticipated in the LOSLR "Vision, Goals and Guiding Principles" (LOSLR, 28 August 2003):

- Criteria and Regulation Plans will incorporate flexible management of levels and flows in recognition of unusual or unexpected conditions affecting the Lake Ontario-St. Lawrence River System.

- Regulation of the Lake Ontario-St. Lawrence River System will be adaptable to reflect the potential for changes in water supply as a result of climate change and variability.
- Criteria and Regulation Plans will incorporate current knowledge, state-of-the-art technology and the flexibility to adapt to future advances in knowledge, science and technology.

These guiding principles were pursued in the Shared Vision planning and modeling processes, and they should continue in some form if the IJC is to achieve the type of long-term, comprehensive evaluation and understanding of tradeoffs among RPOs that is needed. The Shared Vision approach involves an adaptive process. While SVM documentation does not elaborate its links with the larger field of Adaptive Management, the remainder of this section briefly indicates how Adaptive Management might be used to inform future water level and flow regulation.

Brief Description of Adaptive Management Concepts and Alternatives

A recent NRC report defines adaptive management as "…a strategy that aims to create flexible resource management policies that can be adjusted as project outcomes are better understood and as stakeholder preferences change" (NRC, 2004, p. 13). Adaptive management emerged partly in response to unanticipated environmental variability, changing social objectives, and new scientific knowledge. It also addresses problems of protracted stakeholder conflict, inflexible management institutions, and unmonitored trial-and-error responses to these challenges (Gunderson 1999; Lee, 1999; NRC, 2004; Walters, 1997). Many of these issues apply to Plan 1958D: while deviations yield a practical understanding of adjustments to adverse effects, they lack formal scientific hypothesis testing, monitoring, and stakeholder input. Adaptive management strives to combine scientific analysis and stakeholder input for managing complex water and environmental systems such as the Lake Ontario-St. Lawrence River basin.

There are several approaches to adaptive management. A "passive" approach would focus on monitoring the effects of RPO decisions, along with hydrologic, environmental and social variability, and feed that evidence back to managers and stakeholders for consideration in alternative system adjustments and mitigation. An "active" approach would treat RPO decisions as quasi-experimental choices to test the behavior of the LOSLR system in ways that are hypothesized to reduce adverse impacts and/or increase net benefits. There is an increasing body of peer-reviewed scientific literature on the structure and performance of adaptive management programs in regions ranging from the Columbia River to Glen Canyon Dam, the Upper Mississippi River, and the Everglades–as well as programs in Europe (NRC, 1999, 2004). There are several common components of adaptive management programs (NRC, 2004, p. 16):

- Management objectives and performance indicators that are regularly revisited.
- A model of the system that is managed with feedback effects and exogenous driving-forces, as well as management variables.
- Consideration of a range of management alternatives and hypotheses about their possible outcomes.
- Monitoring and evaluation of outcomes.
- A collaborative structure and processes for stakeholder participation.

The LOSLR Shared Vision process prepares the way for these components of adaptive management through its studies, model development, performance indicators, supporting qualitative inquiry, technical work groups, and stakeholder processes.

Flaws in LOSLR models and studies need to be corrected or alternative models adopted prior to the establishment of an adaptive management program to avoid perpetuating existing problems. This would include representative wetland sampling, additional analysis of sandy shore environments, and replacement of regression models for flow routing and shore recession, along with a comprehensive approach to uncertainty analysis, quality assurance, and documentation. After that, an adaptive management program could progressively fill scientific gaps, analyze uncertainties, build cooperative stakeholder relationships, and improve system performance.

The challenges of implementing an adaptive management in the Lake Ontario-St. Lawrence River basin should not be underestimated (e.g., see Johnson, 1999; Lee, 1999; Walters, 1997). Models would need to be implemented in a manner that facilitates future updating at regular intervals. New models would need to be designed to incorporate environmental and socioeconomic processes currently treated as exogenous to RPO decision making. Water regulation decisions, monitoring effects, data analysis, and feedback via the models to subsequent decision making would need to be institutionalized as integral parts of the LOSLR management system. Strong stakeholder support would be needed to initiate and sustain an adaptive management program. The costs of these programs may be high. An "active" adaptive management could involve annual costs comparable to those of the LOSLR study, while "passive" adaptive management costs would be significantly lower, depending upon the scope of monitoring and management, but also less useful. Practical lessons learned in adaptive management programs in the U.S. and internationally can inform the design of a management program for the Lake Ontario-St. Lawrence River system (e.g., see the Collaborative Adaptive Management Network [*http://www.adaptive-management.net/index.php*]). **Recommendation: In the short-term, identify adaptive management alternatives that build upon the LOSLR studies and models. In the longer term, the IJC should, in collaboration with other scientific and stakeholder organizations in the basin, develop an adaptive management program that would provide a con-**

tinuing scientific basis for improving decisions about regulation plan options.

REFERENCES

BACK 2. St. Lawrence River-Lake Ontario Plan of Study Team. September 1999. Plan of Study for Criteria Review in the Orders of Approval for Regulation of Lake Ontario-St. Lawrence River Levels and Flows.

Clinton Edmonds Associates. 2002. Lake Ontario and St. Lawrence River - Changes in the Institutional Structure and Their Impact on Water Levels, 1950-2001. Report to International Lake Ontario-St. Lawrence River Study Board. University of Ottawa, Institute of the Environment.

Economic Advisory Committee. January 7, 2004. Issues and Findings in the Economic Analysis of Water Level Management Plans: Consensus Report of the Economics Advisory Committee. Typescript.

FEPS 11. Contextual Narrative for "Sediment Budget PI" Lake Ontario and the Upper St. Lawrence River.

FEPS 12. Contextual Narrative for Erosion, Flooding, and Existing Shore Protection PIs Lake Ontario and the Upper St. Lawrence River.

FEPS 13. Contextual Narrative for "Beach Access" and "Barrier Beaches & Dunes PIs" Lake Ontario and the Upper St. Lawrence River.

Gunderson, L. 1999. Resilience, flexibility and adaptive management – antidotes for spurious certitude? Conservation Ecology 3(1): 7. Available on-line at *http://www.consecol.org/vol3/ iss1/art7*.

IERM 4. IERM User's Manual (two versions: a) html version with hyperlinks; and b) text version which is also Attachment B of the descriptive documentation).

International Joint Commission. 1956. Order of Approval for Regulation of Lake Ontario. October 29, 1952, as amended by a supplementary Order dated July 2, 1956. Office consolidation. Reprinted in St. Lawrence River-Lake Ontario Plan of Study Team (1999, Annex G).

Johnson, B. L. 1999. Introduction to the special feature: Adaptive management—Scientifically sound, socially challenged? Conservation Ecology 3(1): 10. Available on-line at *http://www.consecol.org/vol3/iss1/art10/*.

Lake Ontario-St. Lawrence River Study Board (LOSLR). 2005. Options for Managing Lake Ontario and St. Lawrence River Water Levels & Flows: Final Report. July 1, 2005 draft. Copy on file at The National Academies, Water Science and Technology Board, 500 5th St., NW Washington, DC 20001.

Lee, K. N. 1999. Appraising Adaptive Management. Conservation Ecology 3(2): 3 Available on-line at *http://www.consecol.org/vol3/iss2/art3*. Accessed October 2005.

Leger, W. 2005. Position paper on conceptual narratives. Revised draft 22 January 2005; and updated 6/7/05 (LOSLR ftp site).

LOSLR (Lake Ontario-St.Lawrence River Study). 2003. Vision, Goals and Guiding Principles. Available on-line at *http://www.losl.org/about/vision-e.html.*

NRC (National Research Council). 1999. Downstream: Adaptive Management of Glen Canyon Dam and the Colorado River Ecosystem. Washington, DC: National Academies Press.

NRC. 2004. Adaptive Management for Water Resources Project Planning. Washington, DC: National Academies Press.

SAR 3A. Impact of Water Level Regulation on Nearshore Habitat Availability and SAR.

SAR 3B. (Supplement) Impact on dunes and SAR.

Sousounis, P. J. (Ed.). 2002. Special section on the potential impacts of climate change in the Great Lakes Region. Journal of Great Lakes Research 28(4): 493-642

Walters, C. J. 1997. Challenges of Adaptive Management of Riparian and Coastal Ecosystems. Conservation Ecology Available on-line at *http://www.consecol.org/vol1/iss2/art1*. Accessed October 2005.

Appendixes

Appendix A
Statement of Task

The Committee shall perform an independent review of the Lake Ontario/St. Lawrence River reports in the following areas: wetlands science and species at risk, the Flood Erosion and Prediction System (FEPS), the Integrated Ecological Response Models (IERM), and the Shared Vision Model (SVM). The level of emphasis for these various areas shall be approximately as follows: wetlands 15%, species at risk 15%, FEPS 10%, IERM 20%, SVM 40%, and reflect the International Joint Commission's (IJC) determination of its priorities in this effort.

The overarching charge shall be to evaluate the appropriateness and sufficiency of the studies and models used to inform decisions related to regulation plan options. Recommendations shall be limited to those deriving from this overarching charge and shall not address management or policy issues.

The Lake Ontario/St. Lawrence River program science, as represented in the reports and model documentation provided (items in articles 11K and 11L), shall be reviewed by the Committee in terms of the degree to which:

(1) the models and reports are sufficient and appropriate to evaluate the various regulation plan options (RPOs) and impacts of changes in water levels and flows;

(2) the studies reflect reasonable scientific methods, assumptions and supported findings;

(3) the models sufficiently and appropriately integrate and display the key information needed for a comprehensive evaluation and understanding of the tradeoffs for selecting among the candidate RPOs.

The review shall be limited to critical evaluation and decision components of the topics listed that relate directly to the Lake Ontario/St. Lawrence River regulation plan options. This requirement shall further be interpreted to restrict the review to the impact of *changing* regulation levels and flows, within the limits that these two factors can be managed using the currently existing control structures and the hydrology/hydraulic characteristics of the system. The review shall neither compare regulation plan options nor provide advice on the preference of one regulation plan option over another, as these actions fall directly within the decision-making responsibilities of the Commission.

Documents to be Reviewed

The following Government owned reports, compute models, and developed data are considered essential to the successful performance under this contract and shall be provided to the Committee by the United States Section of the International Joint Commission.

Wetlands, and Species at Risk – SAR

LOWER ST. LAWRENCE SPECIES AT RISK Sylvain Giguere and Pierre Laporte, Reports 2003-2004, & Mar. 2005 pending.

SELECTED PERFORMANCE INDICATORS OF THE ENVIRONMENTAL TECHNICAL WORKING GROUP (LOWER ST. LAWRENCE). Jean Morin *et al.*, October 2004.

EVALUATION OF THE EFFECT OF LAKE ONTARIO WATER LEVEL REGULATION ON WETLAND PLANT COMMUNITY ABUNDANCE AND DISTRIBUTION Joel Ingram, Nancy Patterson *et al.*, Canadian Wildlife Service, Environment Canada, Ontario Region

MODELS FOR SUBMERGED VEGETATION AND RELATED ENVIRONMENTAL CHANGES INDUCED BY DISCHARGE (WATER LEVEL) VARIATIONS IN THE ST. LAWRENCE RIVER (QUÉBEC) Christiane Hudon, Pierre Gagnon *et al.* St. Lawrence Centre, Environment Canada, Montreal, December 2003

HISTORICAL CHANGES IN HERBACEOUS WETLAND DISTRIBUTION AND BIOMASS : EFFECTS OF HYDROLOGY ON FAUNAL HABITATS IN LAKE ST. PIERRE (ST. LAWRENCE RIVER, QUEBEC, CANADA) Christiane Hudon, Pierre Gagnon et al.,*Environment Canada, St. Lawrence Centre, Montreal*

Context reference already published *Can. J. Fish. Aquat. Sci.* 2004
Hudon, Christiane, SHIFT IN WETLAND COMPOSITION AND BIOMASS FOLLOWING LOW-LEVEL EPISODES IN THE ST. LAWRENCE RIVER: LOOKING INTO THE FUTURE

RELATIONSHIPS BETWEEN LAKE ONTARIO WATER LEVELS AND WETLANDS, U.S. YEAR 3 REPORT FOR THE WETLAND HABITAT QUANTITY PERFORMANCE INDICATOR STUDY. Douglas A. Wilcox, Kurt P. Kowalski, and Martha L. Carlson. (200 page report: 45 pages of text, tables and figures, remainder is appendices with detailed tabular plant quadrat sampling results. Analysis includes all 32 study sites, including data from 16 CND sites).

EVALUATION OF WATER LEVEL REGULATION INFLUENCES ON LAKE ONTARIO AND UPPER ST. LAWRENCE RIVER COASTAL WETLAND PLANT COMMUNITIES —FINAL PROJECT REPORT Douglas Wilcox, Joel Ingram, Kurt Kowalski Martha Carlson, Greg Grabas, Krista Holmes and Nancy Patterson. 100+ pages, Multi-year, binational report on all wetland work completed during the study.

Giguère, S. et P. Laporte (2003) LES ESPÈCES EN PÉRIL INFLUENCÉES PAR LA GESTION DES NIVEAUX D"EAU DU SAINT-LAURENT FLUVIAL (CORNWALL — TROIS RIVIÈRES). Rapport annuel déposé à la CMI (2002-2003). Environnement Canada, Région du Québec, Service canadien de la faune, 179 p.

Giguère, S, J. Morin, P. Laporte and Mingelbier, M. (2005) Évaluation des impacts des fluctuations hydrologiques sur les espèces en péril. Tronçon fluvial du Saint-Laurent (Cornwall — Pointe-du-Lac). Rapport final déposé à CMI (2002 - 2005). Environnement Canada, Région du Québec, Service canadien de la faune, 100 p. + appendix (litterature review = 100 p.)

Species At Risk Performance Indicator Documents

(aa) Black Tern
(ab) Least Bittern
(ac) Lower St. Lawrence River Bridle Shiner
(ad) Lower St. Lawrence River Eastern Sand Darter
(ae) Lower St. Lawrence Softshell Turtle
(af) Lake Ontario Meadow Marsh
(ag) King Rail and Yellow Rail

The following documents will become available at the end of March 2005 or shortly thereafter:

(baa) Lower river wetlands report (vegetation PI)

(bb) Least Bittern (Lake Ontario) and Least bittern (Lower River, Lake St. Louis to Trois Rivières) reproductive index

(bc) Fish SARs report (supplement to overall SAR report)

(bd) Bird SARs report supplement to overall SAR report

(be) Black Tern Reproductive Index. Drolet, Ingram and DesGranges.

(bf) Least Bittern Reproductive Index. Giguèère, Ingram, Drolet, Des-Granges & Laporte

(bg) Lower St Lawrence River SAR 2002-2003 Report (explains which species were looked at, which species were selected to focus on, and reasons leading to these choices) (in French)

(bh) 2003-2004 Report, (includes literature reviews for all selected species, reasons why some of these species were discarded from the study, all work that supports developed performance indicators) (in French)

(bi) SAR report for Lake Ontario and the upper St. Lawrence River (will include discussion of all work over the last 2 years; including basic information on birds and fish, which we expect will be detailed in additional reports (in French)

(bj) General introduction and overview of ETWG study and ETWG Contextual Narrative

***IERM** (Integrated Ecological Response Model)*

A version of the IERM model (to enable reviews to gain hands-on experience, and see how information is displayed).

Summary Documentation and Users' Manual for IERM (contains a listing of all key performance indicators and the Plan Formulation and Evaluation Group's template).

FEPS *(Flood Erosion Prediction System)*

A version of the FEPS model (to enable reviewers to gain hands-on experience and see how information is displayed.)

Automated Lake- wide Flooding Predictions and Economic Damages on Lake Ontario Peter J. Zuzek, MES, P.Geo. and Dr. Robert B. Nairn, P.Eng. (10 pages)

Automated Lake- wide Erosion Predictions and Economic Damages on Lake

Ontario Dr. Robert B. Nairn, P.Eng. and Peter J. Zuzek, (10 pages)

SVM *(Shared Vision Model)*

A version of the SVM (to enable reviewers to gain hands-on experience and see how the information is displayed.)

Appendix B
Documents and Models for Review and Background

SHARED VISION MODEL

For Review

SVM 1 Evaluation Process Overview

SVM 2 Shared Vision Model For The Design and Evaluation of Alternative Regulation of Lake Ontario Releases Into The St. Lawrence River [zip file].

SVM 3 Shared Vision Model for the Design and Evaluation of Alternative Regulation of Lake Ontario Releases into the St. Lawrence River. Documentation Report. February 15, 2005.

For Background

SVM 4 Preliminary Criteria and Metrics for the Plan Formulation and Evaluation Group (PFEG).

SVM 2 Contextual Narratives for evaluating the work of the technical working groups (embedded in the Shared Vision Model).

SVM 6 SVM Informal Decision Tool Program

FLOOD EROSION PREDICTION SYSTEM MODEL

For Review

FEPS 1 Zuzek, P.J., and R. B. Nairn. n.d. Automated Lake wide Flooding Predictions and Economic Damages on LakeOntario.

FEPS 2 Nairn, R.B., and P.J. Zuzek. n.d. Automated Lake wide Erosion Predictions and Economic Damages on LakeOntario.

For Background

FEPS 3 FEPS Model

FEPS 6 Memorandum from Zuzek, P.J., nad R. Roblin (Baird & Associates) to IJC Study Participants re: Integration of Beach User Economics into the Shared Vision Model. April 7, 2004.

FEPS 7 Baird & Associates. March 2004. Erosion Performance Indicator: Methodology and Shared Vision Model Application. Prepared for I.J.C. Plan Formulation and Evaluation Group.

FEPS 8 Baird & Associates. February 2005. Flooding Performance Indicator: Methodology and Shared Vision Model Application. Prepared for IJC Plan Formulation and Evaluation Group. Oakville, Ontario: Baird & Associates.

FEPS 9 Baird & Associates. February 2004. Shore Protection Performance Indicator: Methodology and Shared Vision Model Application. Oakville, Ontario: Baird & Associates.

FEPS 4 Sciremammano, F., and S. Renzetti. 2005. FEPS/Coastal Discussion Paper

FEPS 18 Minutes/Observations from FEPS Independent Technical Review 12/13/2002 and 7/26/2002 notes.

FEPS Contextual Narratives

FEPS 11 Contextual Narrative for "Sediment Budget PI" Lake Ontario and the Upper St. Lawrence River.

FEPS 12 Contextual Narrative for Erosion, Flooding, and Existing Shore Protection PIs Lake Ontario and the Upper St. Lawrence River.

FEPS 13 Contextual Narrative for "Beach Access" and "Barrier Beaches & Dunes PIs" Lake Ontario and the Upper St. Lawrence River.

Performance Indicator Summaries

FEPS 14 Baird & Associates. Erosion Performance Indicator Summary.

FEPS 10 Barrier Beaches and Dunes PI Summary

FEPS 15 Baird & Associates. Flooding Performance Indicator Summary

FEPS 16 Sediment Budget PI Summary

FEPS 17 Baird & Associates. Existing Shoreline Protection Performance Indicator Summary.

FEPS 5 Baird & Associates. Beach Access Performance Indicator Summary

INTEGRATED ECOLOGICAL RESPONSE MODEL

For Review

IERM 1 Descriptive Documentation; Limno-Tech, Inc. April 2005. Draft Development of an Integrated Ecological response Model (IERM) for the Lake Ontario-St Lawrence River Study.

IERM 2 IERM Model (version 4.1.1)

For Background

IERM 3 Development of the IERM for the Lake Ontario – St. Lawrence River Study (March 4, 2005 – Limno-Tech).
IERM 4 IERM User's Manual (two versions: a) html version with hyperlinks; and b) text version which is also Attachment B of the descriptive documentation).
IERM 5 RS – 107 Wetlands Model for the Lower St. Lawrence River [French].
IERM 6 RS – 108 Species at Risk Model for the Lower St. Lawrence River [French].

ENVIRONMENTAL TECHNICAL WORKING GROUP (ETWG)

For Review

ETWG 1 Introduction and Overview of the ETWG
ETWG 2 ETWG Contextual Narrative

For Background

ETWG 3 ETWG Year 3 Technical Report (Not officially reviewed and approved by Study Board; Final draft.).

WETLANDS

For Review

W 1 Wilcox, D. A., K. P. Kowalski, and M. L. Carlson. 2005. Year 3 Report of Relationship between Lake Ontario Water Levels and the Wetlands. Ann Arbor, MI: Great Lakes Science Center, U.S. Geological Survey.
W 2 Evaluation of Water Level Regulation Influences on Lake Ontario and Upper St. Lawrence River Coastal Wetland Plant Communities.

For Background[1]

W 5 Wetland Meadow Marsh PI
W 6 Wetland Meadow Marsh Fact Sheet.
IERM 5 RS -107 Wetlands Model for the Lower St. Lawrence River [French][2]

---

[1] W 3 [models for submerged vegetation, Lower St. Lawrence River] and W 4 [Historic changes in the distribution of wetlands and biomass on Lac St. Pierre] have been removed from review. They have already been peer reviewed and have no impact on determining the Study's regulation plan options.
[2] This a cross reference. Included under IERM since it is the Lower St. Lawrence River IERM modeling component.

SPECIES AT RISK

For Review

SAR 1 Water Fluctuation Impacts on Species At Risk Cornwall to Pointe due Lac [French].

SAR 2 Lake Ontario Species at Risk Supplement (Least Bittern and Black Tern Reproductive Index PIs).

SAR 3A Impact of Water Level Regulation on Nearshore Habitat Availability and SAR.

SAR 3B (Supplement) Impact on dunes and SAR.

SAR 4 Species at Risk Fish Supplement (Year 4 Modeling Group Report).

SAR 5 Lower St. Lawrence Species at Risk Final Report (2005) [French].

For Background

Species at Risk Perform Indicator Background Briefs

SAR 6 Black Tern

SAR 7 Bridle Shiner

SAR 8 Eastern Sand Darter

SAR 9 King Rail

SAR 10 Yellow Rail

SAR 11 Least Bittern

SAR 12 Spiny Soft Shell Turtle

IERM 6 RS – 108 Species at Risk Model for the Lower St. Lawrence River [French][3]

GENERAL BACKGROUND ON THE LAKE ONTARIO ST. LAWRENCE RIVER STUDY

Background Reference Documents[4]

BACK 1 LOSLR Study General Overview Summary

BACK 2 IJC (International Joint Commission). 1999. Plan of Study for Criteria Review in the Orders of Approval for Regulation of Lake Ontario-St. Lawrence River Levels and Flows. Available on-line at *http://www.ijc.org/php/publications/html/pos/pose.html.* Accessed September 2005.

BACK 3 IJC Directive to the Study Board

---

[3] This is a cross reference – included under IERM since it is the Lower St. Lawrence SAR modeling component of the IERM.

[4] Other background resource used is the *http://www.losl.org* website.

BACK 4 Economic Advisors Issues and Findings with respect to economic evaluation of various Study metrics
BACK 5 Year 3 Study Report
BACK 6 Peer Review Triage Strategy (draft)

Appendix C
Acronyms

AARR	average annual recession rate
COSMOS	an erosion model; a process-based numerical model used to calculate coastal sediment transport and morphology
ETWG	environmental technical working group
FEPS	flood-erosion prediction system
GCM	global circulation model
IERM	integrated ecological response model
IJC	International Joint Commission
ISLRBC	International St. Lawrence River Board of Control
LOSLR	Lake Ontario-St. Lawrence River
LO-USL	Lake Ontario-upper St Lawrence
LSL	Lower St. Lawrence
NGO	non-governmental organization
NRC	National Research Council
PI	performance indicator
PIAG	Public Interest Advisory Group
POS	plan of study
QA/QC	quality assurance/quality control
RPO	regulation plan options
RSC	Royal Society of Canada
RSPM	river shore protection model
SAR	species at risk
SAV	submersed aquatic vegetation
STELLA	a computer simulation model written in the STELLA computer language
SVM	shared vision model
SVP	shared vision planning
WAVAD	a wave model; numerical model that generates wave fields in a lake basin with varying winds and bathymetry

Appendix D
Water Science and Technology Board

R. RHODES TRUSSELL, *Chair,* Trussell Technologies, Inc., Pasadena, California
MARY JO BAEDECKER, U.S. Geological Survey (Retired), Vienna, Virginia
JOAN G. EHRENFELD, Rutgers University, New Brunswick, New Jersey
DARA ENTEKHABI, Massachusetts Institute of Technology, Cambridge, Massachusetts
GERALD E. GALLOWAY, Titan Corporation, Reston, Virginia
PETER GLEICK, Pacific Institute for Studies in Development, Environment, and Security, Oakland, California
CHARLES N. HAAS, Drexel University, Philadelphia, Pennsylvania
KAI N. LEE, Williams College, Williamstown, Massachusetts
JAMES K. MITCHELL, Virginia Polytechnic Institute and State University, Blacksburg
CHRISTINE L. MOE, Emory University, Atlanta, Georgia
ROBERT PERCIASEPE, National Audubon Society, New York, New York
LEONARD SHABMAN, Resources for the Future, Washington, DC
KARL K. TUREKIAN, Yale University, New Haven, Connecticut
HAME M. WATT, Independent Consultant, Washington, DC
CLAIRE WELTY, University of Maryland, Baltimore County
JAMES L. WESCOAT, JR., University of Illinois at Urbana-Champaign
GARRET P. WESTERHOFF, Malcolm Pirnie, Inc., White Plains, New York

Staff

STEPHEN D. PARKER, Director
LAUREN E. ALEXANDER, Senior Staff Officer
LAURA J. EHLERS, Senior Staff Officer
JEFFREY W. JACOBS, Senior Staff Officer
STEPHANIE E. JOHNSON, Senior Staff Officer
WILLIAM S. LOGAN, Senior Staff Officer
M. JEANNE AQUILINO, Financial and Administrative Associate
ANITA A. HALL, Administrative Assistant
ELLEN A. DE GUZMAN, Senior Program Associate/Research Associate
DOROTHY K. WEIR, Research Associate

Appendix E
Biographical Information

James L. Wescoat, Jr., *Chair*, is a professor and head of the Department of Landscape Architecture and a faculty affiliate in the Department of Geography at the University of Illinois, Urbana-Champaign. Previously, he was on the faculty of the Department of Geography and member of the Institute of Behavioral Sciences at the University of Colorado, Boulder. His research interests include the historical and cultural geography of water management in North America; and the spatial logic of water law, policies, and institutions. Dr. Wescoat is a member of the Water Science and Technology Board, served on the National Research Council's Committee on the Future of Irrigation in the Face of Competing Demands, and chaired the Committee on Grand Canyon Monitoring and Research. He has conducted comparative analyses of international water policy issues in the Colorado, Indus, and Aral Sea basins; and he recently co-authored a volume with Gilbert F. White titled, *Water for Life: Water Management and Environmental Policy* (Cambridge University Press, 2004). Dr. Wescoat received his M.A. and Ph.D degrees in geography from the University of Chicago.

Patricia Chow-Fraser is a professor at the Department of Biology in McMaster University, Hamilton, Ontario. Her research focused on development of wetland ecosystem-health indicators based on water-quality characteristics, periphyton, zooplankton, benthic macroinvertebrates, aquatic macrophytes and fish obtained from 150 coastal wetlands throughout the U.S. and Canadian shoreline of all five Great Lakes; use of Geographic Information System (GIS) and remote sensing technology to map wetland features (wet meadows, emergent beds, mixed floating-emergent beds, and submergent beds/open water) in coastal wetlands of Georgian Bay and Lake Ontario, and to investigate the spatial relationship between fauna/flora and coastal wetlands, and to determine the impact of land-use alterations and shoreline development on wetland quality; and development of ecological models to determine the aggregate response of coastal wetlands to waterlevel fluctuations, urbanization, and exotic invasions in Lake Ontario. She received her B.Sc. and M.Sc. from the University of Waterloo, and her Ph.D. in zooplankton ecology from the University of Toronto.

Desmond N.D. Hartford is specialist engineer for dam safety risk assessment at British Columbia Hydropower Authority. His areas of expertise and research are in dam safety and catastrophic loss risk management of large dams, hydroelectric facilities, and water resource infrastructure. He is primarily responsible for the development and implementation of risk management solutions for BC Hydro's portfolio of 43 dams. Dr. Hartford also advises dam and water resource agencies across Canada and internationally on dam safety and risk management of wet infrastructure. Dr. Hartford is co-author of Risk and Uncertainty in Dam Safety and is one of the five principal authors of the International Commission on Large Dams (ICOLD) Bulletin 130 on Risk Assessment in Dam Safety Management. Prior to joining BC Hydro Dr. Hartford held senior positions on international infrastructure projects. He is active in training and professional development of engineers in risk management of dams around the world. He received his B.A.I. in civil engineering, M.A. in mathematics, and Ph.D. in soil mechanics from Trinity College in Dublin.

Janet R. Keough is the Acting Director of EPA's Office of Research and Development (ORD) Mid-Continent Ecology Division of the National Health and Environmental Effects Research Laboratory. Her research background is as a wetland ecologist, and she spent most of her research career conducting studies within the Department of Interior, with the U. S. Fish and Wildlife Service, National Biological Service, and U. S. Geological Survey. Dr. Keough's research has focused on the functions of primary producers in wetland habitats and food webs in studies in the Great Lakes and Chesapeake Bay. As Acting Director for this EPA Division, Dr. Keough now guides research to enhance monitoring and assessment programs for large ecosystems, such as Great Lakes nearshore and coastal systems and Great Rivers, and research to improve the predictive capability of risk assessments for chemicals. She serves on the Science Advisory Board for the Minnesota Sea Grant, the Regional Workgroup for the Great Lakes Task Force, and has served on workgroups for NSF in planning for NEON (National Ecological Observing Network). Dr. Keough received her B.S. in biological sciences from Cornell University; her M.A. in plant ecology from Western Michigan University, and her Ph.D. in aquatic ecology from the University of Wisconsin.

Lynn Alison Maguire is an associate professor ath the Nicholas School of Environment and Earth Sciences at Duke University. Her research interests are in conservation biology; applications of decision analysis for environmental decisions and dispute resolutions; public involvement in environmental decision making; and collaborative planning. Dr. Maguire received her A.B. in biology from Harvard University, her M.S. in resource ecology from the University of Michigan, and Ph.D. in ecology from Utah State University.

Joseph C. Makarewicz is a Distinguished Professor in the Department of Environmental Science and Biology at the College at Brockport. His research has focused on Great Lakes research in phytoplankton and zooplankton ecology, the ecology of exotic species and their effect on pelagic food webs, pesticide movement in food webs, and fate and transport of nutrients and herbicides in watersheds. He was recently honored by the Chancellor of the SUNY system for outstanding scholarship and grantsmanship. He is a recipient of the Chandler-Meisner Award for Outstanding Contribution to the Journal of Great Lakes Research by the International Association of Great Lakes Research. Dr. Makarewicz received his Ph.D. from Cornell University

Daene C. McKinney is a professor at the Department of Civil Engineering, University of Texas in Austin. Dr. McKinney's research interests include developing and applying numerical methods for simula-tion, optimization, and uncertainty analysis of environmental and water resource management problems, and the development of laboratory and field experimental techniques for the characterization and remediation of aquifer and groundwater contamination. He is currently engaged in research on: water resource management in the Aral Sea Basin; management and modeling of the Edwards Aquifer in Central Texas; large-scale water-balance computations; expert geographic information systems for water and environmental management; optimal aquifer management and remediation; characterization and remediation of subsurface NAPL contamination; modeling of bioremediation in NAPL contaminated aquifers; and risk-based decision analysis approach for aquifers contaminated with NAPLs. Dr. McKinney received his B.S. from Humboldt State University and his M.S. and Ph.D. from Cornell University.

David H. Moreau is professor in the Departments of City and Regional Planning and Environmental Sciences and Engineering at the University of North Carolina at Chapel Hill. Dr. Moreau teaches water resources planning and regional environmental planning. His research interests include analysis, planning, financing, and evaluation of water resource and related environmental programs. He is actively engaged in water resources planning at the local state, and national levels. He has chaired or served on several NRC committees, most recently as a member of the Committee on Water Quality Improvement for the Pittsburgh Region. Dr. Moreau serves as chairman of the North Carolina Environmental Management Commission, the state's regulatory commission for water quality, air quality, and water allocation. Dr. Moreau received a B.S. and M.S. from Mississippi State University and North Carolina State University, respectively, and a Ph.D. in water resources from Harvard University.

Slobodan P. Simonović is a professor and research chair at the Department of Civil and Environmental Engineering at the Institute for Catastrophic Loss Reduction, The University of Western Ontario. His research encompasses reservoir, flood control, hydropower energy, and operational hydrology. Specifically he is interested in systems modeling; risk and reliability; water resources and environmental systems analysis; computer-based decision support systems development; water resources education and training. He received his B.S. in civil engineering and M.Sc. in interdisciplinary studies, from the University of Belgrade, Belgrade, Yugoslavia and his Ph.D. from University of California, Davis. He is a certified Professional Engineer at Ontario.

Przemyslaw Andy Zielinski is a senior quantitative analyst with Ontario Power Generation. As a senior analyst he develops quantitative and qualitative methodology and analytic tools in assessing operational risk for Ontario Power Generation. He had been previously involved in dam design and development as a dam safety hydrologist and then later as a senior scientist at Ontario Hydro. He was also assistant professor at Warsaw Technical University in Poland. His expertise is in the areas of risk analysis, assessment, and management; applied probability, statistics and stochastic processes; decision making under uncertainty; linear and onlinear optimization; modeling of dynamical systems; and hydrology and water resources management. He presently chairs the Committee on Dam Safety of the International Commision on Large Dams. Dr. Zielinski received his masters in mathematics from the University of Warsaw and his masters in civil engineering and Ph.D. in stochastic hydrology from Warsaw Technical University in Poland.

Staff

Lauren E. Alexander is a senior staff officer with the National Research Council's Water Science and Technology Board. Her research interests include hydro-geomorphic processes and plant diversity in forested wetlands, and she has studied these issues in different coastal plain systems in the United States. Dr. Alexander received her B.S. in applied mathematics and her Masters of Planning in environmental planning from the University of Virginia, and her Ph.D. in landscape ecology from Harvard University. She joined the NRC in 2002.

Ellen A. de Guzman is a research associate with the National Research Council's Water Science and Technology Board. She has worked on a number of studies including Managing the Columbia River, Valuing Ecosystem Services, and Privatization of Water Services in the United States. She co-edits the WSTB newsletter and annual report and manages the WSTB homepage. She received her B.A. degree from the University of the Philippines. She joined the NRC in 1995.